The Map Librarian in the Modern World

Essays in honour of
Walter W. Ristow

Edited by
Helen Wallis and Lothar Zögner

Presented by
the IFLA Section of Geography
and Map Libraries

K · G · Saur
München · New York · London · Paris
1979

K. G. SAUR Verlag KG
Pössenbacherstr. 2b, POB 711009
D-8000 München 71
Federal Republic of Germany
Tel. (089) 79 89 01
Telex 5212067 saur d

ISBN 3-598-10063-9

K. G. SAUR Publishing, Inc.
175 Fifth Avenue
New York, NY 10010
U.S.A.
Tel. (212) 477-2500
Telex 0023/238386

ISBN 0-89664-131-7

CIP-Kurztitelaufnahme der Deutschen Bibliothek

The map librarian in the modern world : essays in honour of Walter W. Ristow / ed. by Helen Wallis and Lothar Zögner. Presented by the IFLA Sect. of Geography and Map Libraries. – München, New York, London, Paris : Saur, 1979.
ISBN 3-598-10063-9

NE: Wallis, Helen [Hrsg.]

CLIVE BINGLEY Ltd. &
K. G. SAUR Ltd.
1-19 New Oxford Street
London WC1A 1NE / UK
Tel. 01-404 4818
Telex 24902 bingle g

K. G. SAUR Editeur S.a.r.l.
38, rue de Bassano
F-75008 Paris
Tel. 72355-18
Telex Iso Bur 630144

Jacket design by Karin Büchner, München
Printed in the Federal Republic of Germany
by grafic + druck GmbH & Co, München
Bound by Thomas-Buchbinderei GmbH, Augsburg

Walter W. Ristow

PREFACE

Dr. Walter Ristow's achievements as scholar-librarian are an inspiration to all of us who believe that the dual career promotes the prosperity of both the world of scholarship and of the libraries who serve that world. For more than thirty years he has found the Library of Congress an environment congenial to his scholarship and to his librarian's hopes and during these years he has helped make the Library a magnet and a resource for other scholars and librarians. His writings in the history of cartography, in map and atlas bibliography and his pioneering efforts in the new world of automation have earned him a place in the history both of American cartographic scholarship and of map librarianship. He has enriched the literature while helping us at the Library discover the promise, the opportunities, the limitations, and the needs of the collections here. His influence as scholar and librarian reach to the farthest boundaries of the world of maps. He has preserved and strengthened the great traditions of the Division of Geography and Maps at the Library of Congress during the decade of his tenure as Chief. We look forward to his continuing wise counsel to the Library and to the world of scholars, as he carries on his increasingly fruitful career. We are all honored by the opportunity to join in honoring him and to cheer on others to follow his example.

Daniel J. Boorstin

The Librarian of Congress

FOREWORD

Map Librarians all over the world were surely present in spirit at the party given to Dr. Walter W. Ristow by the Library of Congress on 27 April 1978 to celebrate his 70th birthday and to honour his 40 years' distinguished service to map librarianship, from which he was then about to retire. Walter's special concern for international co-operation between map librarians has been reflected in his major contribution to the founding and activities of the Geography and Map Libraries Section of IFLA. A group of his colleagues in the Section decided, therefore, to acknowledge this work by dedicating to him a Festschrift compiled in his honour. The papers contained in the volume are intended to cover the many aspects of his professional activities. At the same time they pay tribute to the close friendship and community of interest which their authors have been privileged to share with Walter and Helen Ristow over these years. The volume so dedicated carries fervent good wishes to both for their future happiness.

The editors also would wish to thank all who have contributed to the Festschrift, and likewise Klaus G. Saur who agreed to produce the Festschrift as an IFLA publication. They would mention especially Dr. Daniel J. Boorstin, the Librarian of Congress, whose preface honours Dr. Ristow's achievements as Chief of the Geography and Map Division of the Library of Congress and as a map librarian of international fame.

London, Berlin, June 1978

Helen Wallis

Lothar Zögner

C O N T E N T S

WALTER W. RISTOW

At the end of April 1978, a few days after his seventieth birthday, Walter Ristow retired from the post of Chief of the Geography and Map Division of the Library of Congress. The celebration held at the Library on 27 April honoured his completion of 40 years distinguished service to map librarianship. The fact that the profession has now won full recognition owes much to the example he has set in the fine organization of his Division and his participation in the activities of library and geography associations at home and abroad. Notable among these was the Geography and Map Libraries Subsection (now a Section) of the International Federation of Library Associations (IFLA). His leading role in the founding of the Subsection in 1969 and in guiding its development in its first decade has won general acclaim. That this Festschrift is prepared and published under the auspices of IFLA is therefore particularly appropriate.

Walter William Ristow was born on 20 April 1908 at La Crosse in Wisconsin, the second son and third child in a family of seven. His parents Emil F. and Emily H. Ristow were both children of German immigrants who had settled in La Crosse about 1870. Walter first attended a German Lutheran school in grades 2 to 5, and then was sent to public school, the La Crosse Central High School. Graduating from the school in January 1925, he worked until September 1927 in a supply store and branch office of the Standard Oil Company of Wisconsin. With savings of some 300 dollars he then enrolled at the University of Wisconsin and took up his studies at Madison, at that time the University's single campus.

Walter intended to major in journalism but the geographical instruction which this subject involved aroused his interest, and he changed his course to geography. In 1931 he graduated as a Bachelor of Arts. This was the depth of the economic depression, and he must have been especially grateful to the former lecturer at the University of Wisconsin who now helped him to obtain the post of graduate assistant in the Department of Geology and Geography at Oberlin College, Ohio. He received the degree of M.A. in Oberlin's centennial class of 1933.

A fellowship then took Walter to Clark University, Worcester, Massachusetts, where under the direction of the economic geographer, Professor Clarence F. Jones, he continued his graduate studies. By 1935 he had completed all the requirements for his doctorate except the dissertation. He accepted at 1600 dollars a year a teaching post (which included teaching in the summer session) at Eastern Washington College of Education, Cheney, Washington. In the spring term of 1937 he returned to Clark to complete his dissertation, entitled "Geographic Studies in Orangeburg County, South Carolina", and so obtained the degree of Ph.D. He resumed his teaching at Cheney in the summer session of 1937.

Walter's appointment in September 1937 as Head of the Map Room and Chief of the Map Division at the New York Public Library launched him into the career of map librarian. Like many others in those days (and more recently) he learnt his profession on the job. Founded in 1895 through the amalgamation of three private libraries, the New York Public Library was one of the great American libraries. Its Map Division was established in 1911 and today holds more than 300 000 maps and 6 000 atlases. It serves a public both national and international as well as the New York community. Walter still recalls an encounter with a member of this last group, a lady who asked to see maps of Pomerania. Having provided her with a set of maps, he returned a little later to enquire if she had found what she wanted. "Not yet," she answered. "I'm looking for a name for him", throwing open her coat to reveal a Pomeranian dog.

A week after the attack on Pearl Harbour in December 1942, Walter joined the staff of the New York office of Military Intelligence, U.S. Army, to become head of the geography and map section. He continued this work from December 1942 to August 1944, at the same time directing the Map Division of

New York Public Library, which was then operating on a much reduced scale. In December 1946 Walter accepted the appointment of Assistant Chief of the Map Division of the Library of Congress, and so moved with his family to Washington, D.C. He had married on 14 November 1942, Helen Frances Doerr. Their first son, Walter Richard was born in June 1944, their second son William Whittlesey in February 1947, and their third, Stephen Francis in December 1949.

When the Library of Congress was established in 1800, its holdings included three maps and four atlases. The collection had increased to seven maps and six atlases when the *Catalogue of Books, Maps and Charts Belonging to the Library of the Two Houses of Congress* was published in 1802. The fifty odd maps and an unrecorded number of atlases were destroyed with the rest of the collections when British soldiers burned the Capitol in 1814. To help reconstitute the library, Thomas Jefferson offered his personal collection of books and other materials. These were brought from Monticello to Washington in the spring of 1815. As Jefferson's interests included geography, the Library of Congress had now obtained the nucleus of its map and atlas collections. A second disaster by fire which occurred on 24 December 1851 destroyed some 35 000 books and "nearly all our extensive collections of Maps". Once again the collections had to be built up anew.

The need to establish a separate administrative unit for maps had become evident by the 1850s. Recommendations to this effect were made by Lt. Edward B. Hunt in 1853, Johann Georg Kohl in 1857 and Professor Daniel Coit Gilman of Yale University (in the form of a plea for a Federal Map collection) in 1872. Although these did not bear fruit, the Copyright Act of 1870 and important accessions including the Faden and Rochambeau map collections augmented the cartographic holdings and so strengthened the argument for a Map Division. When the Main Building of the Library of Congress was opened in 1897, a Hall of Maps and Charts was established as a separate unit.

As the map collections grew their rehousing became an urgent necessity. After the second World War, in 1953, they were moved to the Annex Building. In October 1969 they were transferred once again, this time to leased premises on Pickett Street in the suburbs of Alexandria, Va. There they will remain until the James Madison Memorial Building is completed. Walter therefore finished his Map Division Service "out-housed", as it were, but has supervised planning for the return to Capitol Hill. He can proudly claim the Geography and Map Division (so named in 1965) as the largest public map collection in the world, with 38 000 atlases, 7 000 reference books, 3 500 000 printed maps and some 500 three-dimensional relief models. Its preeminence can be established from the *World Directory of Map Collections* (1977), where its entry appears on p.273.

Walter was appointed Associate Chief of the Division under Dr. Arch Gerlach for the period 1966-68 and for various periods was Acting Chief before succeeding in November 1968 as Chief of the Geography and Map Division after Dr. Gerlach had moved to become Chief Geographer of the U.S. Geological Survey (from 1967). Walter followed in a line of distinguished predecessors, from Philip Lee Phillips onwards. Appointed in 1897 as superintendent of the newly created Hall of Maps and Charts, Phillips had already become recognized as a cartographic specialist, and by 1897 had prepared his *List of Maps of America in the Library of Congress*, published in 1901. He went on to provide a documentary record which has become and remains a standard and indispensible work in the literature, his *List of Geographical Atlases in the Library of Congress*, vols. 1-4, published between 1909 and 1920. Col. Lawrence Martin who succeeded Phillips in 1924 was a professional geographer and especially concerned himself with building up the holdings of topographical map series, many of these acquired from United States mapping agencies.

In the second World War Lawrence Martin was transferred to the office of Operational Special Services (O.S.S.) working under A.H. Robinson in its Map Division. Professor Robinson is now the Library of Congress's special consultant in Cartography, and, appropriately, holds the Lawrence Martin professorship of Cartography at the University of Wisconsin-Madison. Robinson and his cartographic team made extensive use of the Geography and Map Division of the Library, and (as Robinson recalls) the first map made by the O.S.S. in the war was, in fact, drawn in the Division. It is true to say of the Geography and Map Division, as of other national map libraries, that the War brought it into the public eye both in official and unofficial circles. No doubt some members of the public first learnt of the Division from the sinister pages of Francis Bonnamy's *Dead Reckoning* (1943), with its intriguing desclaimer "All the persons in this work are creatures of fancy, particularly those in the Map Division of the Library of Congress. Under the direction of Lawrence Martin, that Division is an efficient and distinguished department, with no Miss Fly and nothing misfiled." (The jittery Miss Fly of the Map Division fell victim on page 126 to the treasure-seeking murderer!)

The War had revealed some serious deficiencies in the Library's holdings, especially in its foreign map series. A co-operative map procurement programme was started in 1946 through the Interagency Map and Publication Acquisitions Program and has continued to function through the geographic attachés in various American embassies. Side by side with the programmes of topographic mapping of the official agencies, a National Atlas project was set up at the U.S. Geological Survey, under the direction of Dr. Gerlach, (who was Chief of the Geography and Map Division from 1950). This project bore fruit in the publication of *The National Atlas of the United States of America,* 1970. Designed to meet the needs of "decision-makers" in the realms of government, business, planning and research, it provides a fine example of thematic cartography as applied to economic and social statistics. Dr. Gerlach's leading role emphasizes the close link between the Geography and Map Division and the map-making agencies of the United States. It is also a sign of the times that from 1967 up to his death in 1972 he was Co-ordinator of the Geographic Applications Program for Remote Sensor Data from Aircraft and Space craft.

During Walter's period of service as Associate Chief and then Chief of the Division, the achievements of earlier years have been consolidated. Phillips's work on the compilation of carto-bibliographies was continued by Mrs. Clara Egli Le Gear, who prepared volumes 5, 6 and 7 of the *List of Geographical Atlases in the Library of Congress,* published in 1958, 1963 and 1973. Another useful reference work is her *United States Atlases,* published in two volumes in 1950 and 1953. Early in his career Walter had appreciated the value of the carto-bibliography, producing *The Western Hemisphere ... A list of maps* for the New York Public Library, 1942, and for the Library of Congress in 1956, *Aviation Cartography. A historico-bibliographic study of aeronautical charts* for which the revised and reprinted edition of 1962 is still available. Many carto-bibliographies on special subjects have been prepared by the Geography and Map Division in the last twenty years. Notable examples are Richard W. Stephenson's *Civil War Maps* (1961), Walter Ristow's *Guide to the History of Cartography* (1954, revised edition 1973), and his *Facsimiles of Rare Historical Maps; A List of Reproductions for Sale by various publishers and distributors* (1960; 1978 edition in preparation). One of the most popular reference works of the Division is *A Descriptive List of Treasure Maps and Charts in the Library of Congress*, (first published in 1955; revised edition by Donald H. Wise, 1973), a booklet responding to an interest which the activities of Miss Fly's Division had brought to public notice. The most massive reference work of the Division is the *Bibliography of Cartography*, published by G.K. Hall & Co. in 1973, comprising over 90 000 card references to literature about maps, map-making and map-makers. The range of scholarly work achieved by a generation of officers in the Division is illustrated by the attractive collection of essays aptly titled *À la Carte*, edited by Walter W. Ristow and published in 1972.

Walter's period of service in the Geography and Map Division has been distinguished also by a number of major innovations. The Division has led the way in the development of Machine-readable map cataloguing, although not the first Federal map library in the field. (The U.S. Army Map Service, now the Information Resources Division, Army Topographic Command, had earlier developed a map catalogue on punched cards.) With the help of a grant from the Council on Library Resources, the MARC Map Unit was established within the Library of Congress Processing Section in 1968. It applied first to the cataloguing of new single-sheet maps and map sets to be recorded by title. A report entitled "Machine-Readable Map Cataloguing in the Library of Congress" by Ristow and David K. Carrington, was published in *Special Libraries*, September 1972. In 1973 the Library's Card Division began on a monthly subscription basis the distribution of MARC Map tapes and printed cards. The handbook, *Maps: a MARC Format*; Specifications for Magnetic Tapes containing Catalog records for Maps produced by the Library's Information System office (1976, with addenda) has achieved a world-wide circulation. The development has been in the charge of David Carrington, who was appointed as Co-ordinator in January 1968, and an expert team of graduate geographers with library qualifications has been recruited. Various agencies and institutions overseas have also benefited from the pioneering work of the Geography and Map Division in MARC Maps. Thus, as Hans van de Waal explains in his essay (p.251), the Netherlands took careful account of MARC maps in developing its "Bibliografie van in Nederland verschenen Kaarten", of which the first issue for 1975 was published in 1977.

At the same time that he was directing, in its manifold activities, the highly professional Map Division, Walter did much to promote the profession of map librarianship. In 1951 he instituted the first "Sunrise Seminar" for special projects, henceforward to be held annually. Participants on these projects sort out atlases and sheet maps, selecting duplicates for their sponsoring institutions, and it has become customary for lectures and seminars to be provided. In the past 26 years it is estimated that more than 100 institutions have enrolled and some 350 individuals have attended. Thus many map librarians have benefited from "refresher courses" and from the experience of work in a national map library.

As a young map librarian, Walter was very conscious of the value of professional training and of the opportunities provided by professional associations as a forum for discussion and exchange of ideas. Thus he served as Chairman of the Map Group, New York Chapter of the Special Libraries Association, recently recalling these early days in his paper "The Greening of Map Librarianship", presented to the S.L.A. Meeting at New York in 1977. His appointments to national agencies, geographical and otherwise, have included various important offices. He has Secretary of the American Society of Professional Geographers, 1947-1950, and was re-elected in 1950 for a second three-year term. He represented the Library of

Congress at the First Consultation on Geography of the Pan American Institute of Geography and History, Rio de Janeiro, September 1949. In 1951 he was appointed Chairman of the Middle Atlantic Division of the Association of American Geographers. In 1965 he helped to organize the U.S. Federal Map Users Conference. His work for the American Congress on Surveying and Mapping included membership of the Publications Committee 1960-68, and service as a Cartography Section Councillor, 1960-67. He received from the American Congress on Surveying and Mapping in 1969 an Honorary Citation for his contribution to professional cartography. This was followed by the Congress's Andrew McNally Award in Cartography in 1971, and appointment to Honorary Membership in May 1975. The Special Libraries Association had in 1963 honoured him with the Geography and Map Division Award for Outstanding Achievement. He has attended and read papers at the International Conferences on the History of Cartography at Edinburgh in 1971, Warsaw in 1973, Greenwich in 1975, and his Division acted as host (with the National Archives and the Smithsonian Institution) when this Conference came to Washington, D.C. in 1977 for its most recent meeting.

It seems appropriate to save to the end Walter's major contribution to Map Librarianship in the international sphere, his work for the Geography and Map Libraries Section of IFLA. He was Chairman of the provisional committee which planned the first year's work, 1968-69 and thus secured the founding of the Subsection at the IFLA meeting in August 1969 at Copenhagen. It was his inspiration that the first project taken up by the Subsection should be the compilation of a World Directory of Map Collections. He was director of the project, and he compiled and edited the text from contributions received from libraries all over the world. The publication of the volume in time for the IFLA meeting at Lausanne in September 1976 completed the Subsection's first task. A second edition is being prepared for publication in 1980/1981. Since 1969 Walter has attended all the meetings of the Subsection, which were convened at Moscow in 1970, Budapest 1972, Washington, D.C. 1974, Oslo 1975, Lausanne 1976 and Brussels 1977. At Washington, D.C. in 1974, his Division was host to the Subsection and arranged excellent local excursions. In September 1977, we saw a happy outcome of these activities when under the terms of the New Statutes of IFLA, the Geography and Map Library Section was established in the Special Libraries Division.

This account would not be complete without mention of Helen Ristow, who as Walter's companion on his travels has graced nearly all these occasions abroad and in Washington has been a delightful hostess to visiting map librarians, geographers and historians at their home at McLean, and has looked after the Ladies' (or as we now call it, the Accompanying Persons') Programme. Many will have memories of Walter and Helen at the centre of a group of kindred spirits in some cartographic or historic venue, ranging from Edinburgh Castle to a lakeside café at Lausanne.

Walter Ristow may be accounted one of the most influential figures - perhaps *the* most influential figure - in map librarianship in the United States, and he has won the highest international standing in his field. He has also been a good friend and colleague in the world-wide community of map librarians which has grown up over the last thirty years. This volume pays warm tribute to the achievements of his life and work, and honours the Geography and Map Division of the Library of Congress which he has served so well.

Helen Wallis

LIST OF PUBLICATIONS
by WALTER W. RISTOW

1933

(1) The influence of geography upon the history of Bohemia. In: *Bulletin of the Geographical Society of Philadelphia*, 31, 1933, no. 2, p.73-86.

1936

(2) (With J. Norman Carls) The industrial geography of Seneca Falls, New York. In: *Economic Geography*, 12, 1936, no. 3, p.287-293.

(3) The rise and decline of the indigo industry of South Carolina. [Abstract.] In: *Yearbook of the Association of Pacific Coast Geographers*, 2, 1936, p.32.

1937

(4) Geographic studies in Orangeburg County, South Carolina. Ph.D. Dissertation, Clark University, Worcester, Mass. 1937. 220 p. [Typescript.]

1939

(5) Geographical information please! In: *Journal of Geography*, 38, 1939, no. 8, p.314-318.

1942

(6) The Library map collection. In: *Library Journal*, 67, 1942, no. 12, p.552-555.

(7) [Foreword in] Lawrence, Chester H.: New world horizons; geography for the air age ..., New York: Duell, Sloan and Pearce, 1942, p.9-10.

(8) The western hemisphere, an exhibition held at the New York Public Library. A list of maps of the world and of America from 1492-1942. New York: The New York Public Library, 1942, 27 p.
Also in: *Bulletin of the New York Public Library*, 46, 1942, p.419-444.
Introduction by Walter W. Ristow.

1943

(9) Land patents and manors of seventeenth and eighteenth century New York. [Map.] In: Adams, James Truslow, R.V. Coleman: Atlas of American history.-New York: Charles Scribner's Sons, 1943. Plate 27.

(10) Maps in war time. In: *Education Magazine*, 63, 1943, no. 5, p.273-276.

(11) Maps for global war. [Editorial.] In: *Library Journal*, 68, 1943, no. 8, p.324-325.

(12) Maps: How to make them and read them. In: *Bulletin of the New York Public Library*, 47, 1943, no. 6, p.381-386.
Also issued separately in pamphlet form.

(13) Maps: How to make them and read them. In: *The Journal of Geography*, 42, 1943, no. 7, p.258-265.
Revised and enlarged from (12).

1944

(14) Air age geography: a critical appraisal and bibliography. In: *Journal of Geography*, 43, 1944, no. 9, p.331-343.

(15) The trans-orbal map: a global press map for one world.- New York: Global Press Corp., 1944.
Text by Walter W. Ristow.

(16) Man's attempt to visualize and map his world. In: Renner, George T. and others: Global Geography. New York: Crowell, 1944. Chapter XI (p.188-214).

1945

(17) The human side of maps. In: *Wilson Library Bulletin*, 20, 1945, no. 2, p.151-152.

(18) A survey of world atlases. In: *Library Journal*, 70, 1945, no. 2, p.54-57; no. 3, p.100-103.

(19) (With others) World reference atlases, a survey of current resources. In: *Bulletin of the American Society for Professional Geographers*, 3, 1945, nos. 5-6, p.1-8.

1946

(20) [Review of] Boggs, Samuel W., Dorothy C. Lewis: The classification and cataloging of maps and atlases. New York 1945. In: *Geographical Review*, 36, 1946, no. 2, p.347-348.

(21) American Road maps and guides. In: *The Scientific Monthly*, 62, 1946, no. 5, p.397-406.

(22) Maps in libraries, a bibliographical summary. In: *Library Journal*, 71, 1946, no. 15, p. 1101-1107; 1121-1124.

(23) Thank the wheelman for your road map. In: *Frontiers;* A Magazine of Natural History, 11, 1946, no. 2, p.56-59.

1947

(24) (With others) World reference atlases, a survey of current resources. In: *Special Libraries*, 38, 1947, no. 3, p.67-73; no. 4, p.100-108. Repr. from (19).

(25) Maps of Cloth. In: *The Magazine Antiques*, 51, 1947, no. 4, p.256-258.

1949

(26) (With staff) Maps [Annual Report of Acquisitions, Division of Maps]. In: Library of Congress. *Quarterly Journal of Current Acquisitions*, 6, 1949, no. 4, p.63-74.

(27) The Round Earth on Flat Paper, Review of "The Story of Maps" by Lloyd A. Brown. (Boston 1949.) In: *The Scientific Monthly*, 69, 1949, no. 4, p.277-278.

1950

(28) (With staff) Maps [Annual Report of Acquisitions, Map Division]. In: Library of Congress. *Quarterly Journal of Current Acquisitions*, 7, 1950, no. 4, p.59-68.

(29) [Anonym.] Map Division. - Washington, D.C.: The Library of Congress 1950. 43 p. Multilith. (Library of Congress Departmental and Divisional Manuals No. 15.)

(30) Cartographical Publications of the Library of Congress. In: *Professional Geographer*, N.S. 2, 1950, no. 6, p. 25-31.

1951

(31) Marketing maps of the United States. An annotated list. Washington, D.C.: The Library of Congress, Map Division, 1951. 54 p. Multilith.

(32) [Review of] World Geography of Petroleum. Ed. by the American Geographical Society. Princeton 1950. In: *U.S. Quarterly Book Review*, March, 1951.

(33) The Services and Collections of the Map Division. The Library of Congress. - Washington, D.C.: U.S. Government Printing Office, 1951. 22 p., 17 plates.

(34) Three-dimensional maps, an annotated list of references.- Washington, D.C.: The Library of Congress, Map Division, 1951. 43 p. Multilith.

(35) Some recent regional atlases: a review. In: *Geographical Review*, 41, 1951, no. 3, p.479-484.

(36) (With staff) Maps [Annual Report of Acquisitions, Map Division]. In: Library of Congress. *Quarterly Journal of Current Acquisitions*, 8, 1951, no. 4, p.43-55.

1952

(37) The AAG, the IGU, and the Seventeenth International Geographical Congress. In: *Professional Geographer*, 4, 1952, p.21-24.

(38) (With Marie C. Goodman) Marketing Maps of the United States, an annotated bibliography. 2. (rev.) ed. - Washington, D.C.: The Library of Congress, Map Division, 1952, 100 p. Multilith.

(39) Preview of the Seventeenth Congress of IGU. In: *Journal of Geography*, 51, 1952, no. 5, p.220.

1953

(40) [Review of] Skelton, R.A.: Decorative Printed Maps of the 15th to 18th Centuries. In: *Geographical Review*, 43, 1953, no. 3, p.446-447.

(41) Maps [Annual Report of Acquisition, Map Division]. In: Library of Congress. *Quarterly Journal of Current Acquisitions*, 10, 1953, no. 4, p.227-238.

(42) [Review of] Chapin, Edward L.: A Selected Bibliography of Southern California Maps. Chapel Hill 1953. In: *Surveying and Mapping*, 13, 1953, no. 3, p.379-380.

1954

(43) [Review of] Rapport de la Commission pour la Bibliographie des Cartes Anciennes. Union Géographique Internationale 1952. In: *Geographical Review*, 44, 1954, no. 1, p.161-163.

(44) The bibliographic record of the Seventeenth International Geographical Congress. In: *IGU Newsletter*, 4, 1953, p.9-23; 5, 1954, p.38-40.

(45) (With Clara E. LeGear) A guide to historical cartography. - Washington, D.C.: The Library of Congress, 1954, 19 p. Multilith.

(46) Cartographic exhibits. In: *Surveying and Mapping*, 14, 1954, no. 1, p.18-25.

(47) [Review of] Atlas över Sverige. Stockholm 1953. In: *Surveying and Mapping*, 14, 1954, no. 2, p.228.

(48) Fiftieth anniversary meeting of the Association of American Geographers. In: *Surveying and Mapping*, 14, 1954, no. 2. p.251.

(49) Maps for extra motoring pleasure. In: *Fine Cars* [Ford Motor Co. publication], 1, 1954, no. 4. p.10-12.

(50) [Review of] Portfolio of United States Census maps 1950. In: *Surveying and Mapping*, 14, 1954, no. 3, p.372.

(51) Maps [Annual report of acquisitions for the Map Division]. In: Library of Congress. *Quarterly Journal of Current Acquisitions*, 11, 1954, no. 4, p.239-249.

1955

(52) Maps [Annual Report of Acquisitions, Map Division]. In: Library of Congress. *Quarterly Journal of Current Acquisitions*, 12, 1955, no. 4, p.217-226.

(53) What About Maps? In: *Library Trends*, 4, 1955, no. 2, p.123-139.

1956

(54) American Council of Learned Societies. In: *Professional Geographer*, 8, 1956, no. 2, p.27-28.

(55) Bibliographie Cartographique Internationale. In: *Surveying and Mapping*, 16, 1956, p.395.
Note about 1954 B.C.I.

(56) One hundred topographic maps illustrating specified physiographic features. In: *Surveying and Mapping*, 16, 1956, no. 1, p.54.

(57) [Review of] Lindberg, Otto G.: My Story. Convent Station, N.J. 1955. In: *Surveying and Mapping*, 16, 1956, p.307-308.

(58) Committee to Select Topographic Quadrangles Illustrating Cultural Geography, Division of Earth Sciences NAS-NRC. (Chairman: Walter W. Ristow.) Rural settlement patterns in the United States as illustrated on one hundred topographic quadrangle maps. - Washington, D.C.: National Academy of Sciences - National Research Council, 1956. 52 p., processed. (Pub. 380.)

(59) Maps [Annual Report of Acquisitions, Map Division, for report year 1956]. In: Library of Congress. *Quarterly Journal of Current Acquisitions*, 13, 1956, p.213-222.

(60) Aviation Cartography. - Washington, D.C. Library of Congress, 1956. 114 p.

(61) [Review of] Powell, Louis H.: Globes and Spherical Maps. Saint Paul, Minn. 1956. In: *Surveying and Mapping*, 16, 1956, p.500-501.

(62) Foreign maps. In: *Surveying and Mapping*, 16, 1956, p.527.

1957

(63) [Review of] Kosack, Hans-Peter, Karl-Heinz Meine: Die Kartographie, 1943-1954. Lahr 1955. In: *Geographical Review*, 47, 1957, p.151-153.

(64) American Council of Learned Societies. In: *Professional Geographer*, 9, 1957, p.32-33.

(65) [Review of] Lobeck, Armin K.: Things Maps Don't Tell us. London, 1956. In: *Scientific Monthly*, 84, 1957, p.159-160.

(66) Visualizing and mapping the world. In: Miller, E. Willard, George T. Renner a.o.: Global Geography. Rev. ed. - New York: Crowell 1957, Chapter 12 (p.171-194).

(67) Aviation Cartography, a historico-bibliographic study of aeronautical charts. - St. Louis: Aeronautical Chart and Information Center, 1957, 115 p. Repr. from (60).

(68) Captain John Smith's Map of Virginia. Text to accompany facsimile edition of the map in the collections of the Library of Congress. - Washington, D.C.: Library of Congress 1957.

(69) U.S. contributions to international organizations. In: *Surveying and Mapping*, 17, 1957, p.182.

(70) [Review of] The story of maps. Issued by General Drafting Co. In: *Surveying and Mapping*, 17, 1957, no. 2, p.218.
Also printed, in error, in 17, 1957, no. 3.

(71) Recent maps of Antarctica. In: *Surveying and Mapping*, 17, 1957, p.232.

(72) Lists of published maps. In: *Surveying and Mapping*, 17, 1957, p.232.

(73) Maps [Annual report on acquisitions, Library of Congress, Map Division]. In: Library of Congress. *Quarterly Journal of Current Acquisitions*, 14, 1957, p.189-198.

(74) Journalistic cartography. In: *Surveying and Mapping*, 17, 1957, p.369-390. Repr. in: The Bobbs-Merill Reprint Series in Geography. Indianapolis 1968.

1958

(75) [Review of] Pattison, William D.: Beginnings of the American Rectangular Land Survey System. Chicago 1957. In: *Surveying and Mapping*, 18, 1958, p.94.

(76) Bibliotheca Cartographica, [note about a new serial publication]. In: *Surveying and Mapping*, 18, 1958, p.128.

(77) Marketing Maps of the United States, an Annotated Bibliography. 3rd (rev.) ed.- Washington, D.C.: Library of Congress, 1958. 147 p.

(78) [Review of] Pangborn, Mark W.: Earth for the Layman. Washington, D.C. 1957. In: *Surveying and Mapping*, 18, 1958, p.233.

(79) Maps of distinction. In: *Surveying and Mapping*, 18, 1958, p.445-448.

(80) [Review of] Cumming, William P.: The Southeast in Early Maps. Princeton, 1958. In: *William and Mary Quarterly*, 15, 1958, p.535-537.

(81) (With Clara E. LeGear) The Melville Eastham Gift of Atlases. In: Library of Congress. *Quarterly Journal of Current Acquisitions*, 15, 1958, p.219-227.

(82) (With Catherine I. Bahn) Maps. [Annual Report on acquisitions.] In: Library of Congress. *Quarterly Journal of Current Acquisitions*, 15, 1958, p.235-243.

(83) (Anonym) [Obituary for George D. Hubbard.] In: *Professional Geographer*, 10, 1958, p.43-44.

(84) American Council of Learned Societies. (Annual report of AAG Delegate to ACLS.) In: *Professional Geographer*, 10, 1958, p.29-31.

1959

(85) Aeronautical charts. [Note on several German publications.] In: *Surveying and Mapping*, 19, 1959, p.54.

(86) (With Clara E. LeGear) Professional Geodetic and Cartographic Literature in the Soviet Union. In: *Surveying and Mapping*, 19, 1959, p.91-93.
Translation of: Deumlich, F.: Die geodätische und kartographische Fachliteratur in der Sowjetunion, in Vermessungstechnik, 5, 1957, 11, p.263-264.

(87) Second coastal geography conference. (April 6-9, 1959.) In: *Surveying and Mapping*, 19, 1959, p.82.

(88) [Review of] Kartographische Studien. Haack-Festschrift. Hermann Lautensach and Hans-Richard Fischer. Hrsg. von Petermanns Geographische Mitteilungen, Ergänzungsheft Nr. 264, 1957. In: *Geographical Review*, 49, 1959, no. 2, p.293-295.

(89) Imago Mundi. [Brief note on appointment of new Editor for Imago Mundi.] In: *Surveying and Mapping*, 19, 1959, p.49.

(90) Humboldt Centenary [Announcement of centenary to be held in Berlin May 18-19, 1959]. In: *Surveying and Mapping*, 19, 1959, p.90.

(91) [Review of cartography issue of Dimensions, Vicksburg, Mich., Lee Paper Co. 1958.] In: *Surveying and Mapping*, 19, 1959, p.90.

(92) International Congress for the History of Discoveries [to be held in Lisbon, Portugal, Aug. 7-15, 1960]. In: *Surveying and Mapping*, 19, 1959, p.86.

(93) (With Catherine I. Bahn) Maps [Annual Report of Map Division.] In: Library of Congress. *Quarterly Journal of Current Acquisitions*, 16, 1959, p.173-182.

(94) Map supplement to the Annals of the Association of American Geographers.[Note] In: *Surveying and Mapping*, 19, 1959, p.284.

(95) Fifty years of aviation cartography. In: *Surveying and Mapping*, 19, 1959, no. 3, p.377.

(96) [Review of] Rand McNally Bible Atlas, 1956; Rand McNally Atlas of World History, 1957; Atlas of the Early Christian World, 1958. In: *Geographical Review*, 49, 1959, p.595-597.

(97) Fifty years of aviation cartography. [Note about an exhibit which opened in Munich, Germany on Sept. 25, 1959.] In: *Surveying and Mapping*, 19, 1959, p.511, 521.

(98) Facsimiles of rare historical maps available for sale. - Washington, D.C.: Library of Congress, Map Division, Jan. 1960. 6 p.
Repr. May 1960.

(99) [Review of] Brun, Christian: Guide to the manuscript maps in the William L. Clements Library. Ann Arbor 1959. In: *American Archivist*, 23, 1960, p.75-76.

(100) (With Clara E. LeGear) A guide to historical cartography. 2nd rev. ed. - Washington, D.C.: Library of Congress, 1960. 22 p.
Repr. 1962.

(101) (With Catherine I. Bahn) Maps [Annual Report on Acquisitions, Map Division.] In: Library of Congress. *Quarterly Journal of Current Acquisitions*, 17, 1960, p.233-241.

(102) Augustine Herrman's Map of Virginia and Maryland. In: Library of Congress. *Quarterly Journal of Current Acquisitions*, 17, 1960, p.221-226.
Repr. in: *Surveying and Mapping*, 21, 1961, p.57-62.

(103) Aeronautical charts - fifty years of evolution. In: Association of American Geographers. *Annals*, 50, 1960, p.344.

(104) Geographical Journal. In: *Surveying and Mapping*, 20, 1960, p.276, 278.

(105) Fifth century of the death of Prince Henry the Navigator. In: *Surveying and Mapping*, 20, 1960, p.276.

(106) Aviation Cartography, a historico-bibliographic study of aeronautical charts. 2nd (rev.) ed. - Washington, D.C.: The Library of Congress, 1960. 245 p.
Repr. 1962.

(107) Weather and climate maps. Preliminary draft of section of Cartographic Research Guide. In: Special Libraries Association. Geography and Map Division. *Bulletin*, no. 42, 1960, p.24-38.

(108) [Review of] Meer, Frederic van der: Atlas of Western Civilization. 2nd, rev. ed. Amsterdam 1960. In: Special Libraries Association. Geography and Map Division. *Bulletin*, no. 42, 1960, p.49-50.

1961

(109) John Melish, Geographer and Map Publisher. In: Association of American Geographers. *Annals*, 51, 1961, p.420.

(110) (With Catherine I. Bahn) Maps [Annual Report of Acquisitions for Map Division]. In: Library of Congress. *Quarterly Journal of Current Acquisitions*, 18, 1961, p.207-217.

(111) A survey of the roads of the United States of America, 1789, by Christopher Colles. - Cambridge, Mass.: Belknap Press of Harvard University Press, 1961. 227 p. ill., maps.
Edited by Walter W. Ristow.
Part one: Christopher Colles, early American engineer and cartographer. By Walter W. Ristow.

1962

(112) [Review of] Lunny, Robert M.: Early maps of North America. Newark, N.J.,1961. In: *Surveying and Mapping*, 22, 1962, p.116.

(113) [Review of] International Yearbook of Cartography. 1, 1961. In: *Surveying and Mapping*, 22, 1962, p.116.

(114) World atlases, a listing of the currently available American works recommended "as best suited for library reference use". In: *Library Journal*, 87, 1962, p.1553-1554.

(115) (With Catherine I. Bahn) Maps [Map Division annual report on acquisitions]. In: Library of Congress. *Quarterly Journal of Current Acquisitions*, 19, 1962. p.181-191.

(116) John Melish and his map of the United States. In: Libraryof Congress. *Quarterly Journal of Current Acquisitions*, 19, 1962, p.159-178.

(117) Early American atlases. In: Association of American Geographers. *Annals*, 52, 1962, p.375.

(118) Early American atlases. In: *Surveying and Mapping*, 22, 1962, p.569-574.

(119) The Juan de la Cruz Map of South America, 1775. In: Festschrift Clarence F. Jones. Ed. by Merle C. Prunty, Jr. Evanston, Illinois, 1962. Chapter 1. (Northwestern University Studies in Geography, No. 6.)
Also in Spanish: El mapa de Sud América de Juan de la Cruz y Olmedilla. In: *Revista Chilena de Historia y Geografia*, 132, 1964, p.94-109.

1963

(120) [Review of] Schweizerischer Mittelschulatlas. Hrsg. v. Eduard Imhof. 13. Aufl. Zürich 1962. In: *Surveying and Mapping*, 23, 1963, p.124.

(121) Maps and map making. In: *Grolier Encyclopedia*, 13, 1963. p.205-209.

(122) Map Collector's Series. [Note] In: *Surveying and Mapping*, 23, 1963, p.409.

(123) (With Catherine I. Bahn) Map [Annual report on acquisitions, Map Division]. In: Library of Congress. *Quarterly Journal of Current Acquisitions*, 20, 1963, p. 225-234.

(124) About the program. In: Association of American Geographers, 59th *Annual Convention*. Denver, Colo., Sept. 1-5, 1963. Program: p.3-4.

(125) Historical Cartography in the United States, 1959-1963. In: *Imago Mundi*, 17, 1963, p.106-114.
Bibliography with 121 items.

1964

(126) Fortieth anniversary of [death of] Philip Lee Phillips, 1856-1924. In: Library of Congress. *Information Bulletin*, 23, 1964, p.5-6.

(127) [Review of] National Geographic Atlas of the World. Washington 1963. In: *Geographical Review*, 54, 1964, p.298-300.

(128) [Foreword to] A Descriptive List of Treasure Maps and Charts in the Library of Congress. Comp. by Richard S. Ladd. - Washington, D.C.: Library of Congress, 1964.

(129) [Review of] Warntz, William: Geography now and then, New York 1964. In: SLA Geography & Map Division. *Bulletin*, no. 57, 1964, p.22.

(130) Simeon DeWitt, New York State's first professional geographer. [Abstract] In: Association of American Geographer. *Annals*, 54, 1964, p.435.

(131) (With Catherine I. Bahn) Maps [Annual Report on Acquisitions, Map Division.] In: *Quarterly Journal of the Library of Congress*, 21, 1964, p.273-284.

(132) Remarks. In: Federal Map Users Conference on National Topographic Program, Oct. 5-6, 1964. Washington, D.C., [1964] p.67-71.

(133) A Half Century of Oil-Company Road Maps. In: *Surveying and Mapping*, 24, 1964, p.617-637.

(134) Maps on Exhibit. [Note] In: *Surveying and Mapping*, 24, 1964, p.670.

(135) Three-dimensional maps, an annotated list of references relating to the construction and use of terrain models. 2nd rev. ed. - Washington, D.C.: Library of Congress 1964. 38 p.

(136) [Review of] A new and correct map of the United States of America by Abel Buell. Facs., pub. by New Jersey Historical Society 1963. In: *Imago Mundi*, 18, 1964, p.98-99.

1965

(137) (With Catherine I. Bahn) Maps [Annual Report on Acquisitions, Map Division]. In: *Quarterly Journal of the Library of Congress*, 22, 1965, p.216-225.

(138) [Review of] Bagrow, Leo; R.A. Skelton: History of Cartography. London 1964. In: *Pacific Northwest Quarterly*, 56, 1965, p.95-96.

(139) Werner projection used in International Cardiology Foundation emblem. In: *Surveying and Mapping*, 25, 1965, p.480.

(140) Nineteenth century cadastral maps in Ohio. In: Bibliographical Society of America. *Papers*, 59, 1965, p.306-315.
Abstract in: Association of American Geographers. *Annals*, 55, 1965, p.643.

1966

(141) [Review of] Facsimile edition of Lucan Janszoon Waghenaer's Spieghel der Zeevaerdt, 1584-1585. New York 1964. In: *Geographical Review*, 56, 1966, p.311-312.

(142) In memorian [sic]. (Obituary notes on Lloyd A. Brown, Arthur Bolton Carlson, Carl H. Mapes and Carl I. Wheat.) In: SLA Geography and Map Division. *Bulletin*, no. 65, 1966, p.11-14.

(143) Deaths of distinguished historical cartographers. In: *Surveying and Mapping*, 26, 1966, p.479-480.

(144) Facsimiles of rare historical maps. A list of reproductions for sale by various publishers and distributors. 2nd rev. ed. - Washington, D.C.: Library of Congress, Geography and Map Division, 1966. 12 p.
Repr. 1967.

(145) A covey of names (Birds and geographical names). In: Birds in our lives. Ed. by Alfred Stefferud. Washington, D.C.: U.S. Department of the Interior, Bureau of Sport Fisheries and Wildlife, Fish and Wildlife Service, 1966, p.68-77.
Repr. in: *Surveying and Mapping*, 30, 1970, p.419-426.

(146) Maps [Annual Report on Acquisitions]. In: *Quarterly Journal of the Library of Congress*, 23, 1966, p.231-242.
Includes data on early maps of Pennsylvania and Virginia.

(147) Alfred T. Andreas and his Minnesota Atlas. In: *Minnesota History*, 40, 1966, no. 3, p.120-129.

(148) Maps and globes. In: New Book of Knowledge. New York: Grolier 1966. p.80-87.

(149) State maps of the Southeast to 1833. In: *The Southeastern Geographer*, 6, 1966, p.33-40.

(150) Chronicles: United States. In: *Imago Mundi*, 20, 1966, p.90-94.

1967

(151) Seventeenth century wall maps of America and Africa. In: *Quarterly Journal of the Library of Congress*, 24, 1967, p.2-17.

(152) Map libraries and automation, a bibliography. Comp. by Geography and Map Division's Committee on Automation. (Chairman: Walter W. Ristow) In: Special Libraries Association. Geography and Map Division. *Bulletin*, no. 66, 1966, p.14-18; no. 67, 1967, p.14-19.

(153) Recent facsimile maps and atlases. In: *Quarterly Journal of the Library of Congress*, 24, 1967, p.213-229.

(154) The emergence of maps in Libraries. In: *Special Libraries*,58, 1967, p.400-419.

(155) [Review of]Thomson, Don W.: Men and meridians: the history of surveying and mapping in Canada. Ottawa 1966. In: *Minnesota History*, 1967, p.355-356.

(156) Map librarianship; one of the underdeveloped areas of library education. In: *Library Journal*, 92, 1967, p.3610-3614.

(157) [Review of]Day, Sir Archibald: The Admiralty Hydrographic Service 1795-1919. London 1967. In: *Surveying and Mapping*, 28, 1968, p.332.

(158) Remarks. In: Federal Map Coordination Conference. *Proceedings*. Washington, D.C.: U.S.G.S. 1967, p.43-45.

1968

(159) Simeon De Witt, Pioneer American Cartographer. In: Kartengeschichte und Kartenbearbeitung. Festschrift zum 80. Geburtstag von Wilhelm Bonacker. Hrsg. von Karl-Heinz Meine. Bad Godesberg: Kirschbaum-Verlag 1968, p.103-114.
Repr. in: *Canadian Cartographer*, 5, 1968, p.90-107.
Also in: *Surveying and Mapping*, 30, 1970, p.239-255.

(160) United States Fire Insurance und Underwriters Maps 1852-1968. In: *Quarterly Journal of the Library of Congress*, 25, 1968, p.194-218.

(161) New maps from old: trends in cartographic facsimile publishing, In: *Canadian Cartographer*, 5, 1968, p.1-17. Repr. with a different title and slight revision, from (153).

(162) [Review of] Waters, D.W.: The Rutters of the Sea. New Haven 1967. In: *American Historical Review*, 1968, p.148-149.

(163) [Review of] Crone, G.R.: Maps and their makers. New York 1966. In: *Professional Geographer*, 20, 1968, p.450-451.

(164) (Assisted by Mary E. Graziani) Facsimiles of rare historical maps. A list of reproductions for sale by various publishers and distributors. 3rd ed., rev. and enlarged. - Washington, D.C.: Library of Congress, 1968 [i.e. 1969], 20 p.

1969

(165) Erwin Raisz 1893-1968. In: SLA Geography and Map Division. *Bulletin*, no. 75, 1969, p.9-10.

(166) Erwin Raisz, 1893-1968. In: *Canadian Cartographer*, 6, 1969, p.59-77.
Adapted from notice in SLA Geography and Map Division. *Bulletin*, 75, 1969. See no. (165).

(167) The map publishing career of Robert Pearsall Smith. In: *Quarterly Journal of the Library of Congress*, 26, 1969, p.170-196.

(168) (With David K. Carrington) The Library of Congress computerized map cataloging project. In: *INSPEL*, 4, 1969, p.74-79.
Also in: Special Libraries - Worldwide. Ed. by G. Reichardt. München/Pullach: Verlag Dokumentation 1974, p.31-37. (IFLA Publications. 1.)
Paper presented at the meeting of the Geography and Map Libraries Subsection, IFLA. Kopenhagen, August 1969.

(169) Geography and Map Libraries Subsection organized within the Special Libraries Section of IFLA. In: SLA Geography and Map Division. *Bulletin*, 78, 1969, p.43-44.

(170) Kartensammlung und Kartenbibliothekarswesen in den Vereinigten Staaten. Zwei Berichte von Walter W. Ristow, übersetzt von J. Neumann. Bonn-Bad Godesberg 1969. (Kartensammlung und Kartendokumentation. 6.)
[1.] Das Problem der Karten in den Bibliotheken, p.3-34. Also in English, see (156).
[2.] Das Kartenbibliothekarswesen, p.35-45. Also in English, see (154).

1970

(171) U.S. fire insurance maps, 1852-1968. In: *Surveying and Mapping*, 30, 1970, p.19-41.
Repr. with slight modifications from (160).

(172) Automated map cataloging at the Library of Congress - background and development. In: Association of Canadian Map Libraries. Third Annual Conference, June 4-6, 1969. *Proceedings*. Quebec [1970], p.37-44.

(173) Geography and Map Libraries Subsection, International Federation of Library Associations. In: *International Library Review*, 1970, no. 2, p.387-389.

(174) (With David K. Carrington) The Library of Congress computerized map cataloging project. In: *International Library Review*, 1970, no. 2, p.391-397.

(175) The Haile Selassie Map of Ethiopia. In: *Quarterly Journal of the Library of Congress*, 27, 1970, p.257-266.

(176) IFLA Geography and Map Libraries Subsection. Report on Sessions at Moscow IFLA Conference. Sept. 1-5, 1970. In: *INSPEL*, 5, 1970, p.55-60.

(177) Theatrum Orbis Terrarum 1570-1970. In: *Quarterly Journal of the Library of Congress*, 27, 1970, no. 4, p.316-331, ill.

(178) International map librarianship, IFLA (Moscow), 1970. In: SLA Geography and Map Division. *Bulletin*, no. 82, 1970, p.42-43.

1971

(179) Death of R.A. Skelton. In: SLA Geography and Map Division. *Bulletin*, no. 83, 1971, p.52-53.

(180) (Assisted by Mary E. Graziani) Facsimiles of rare historical maps, a list of reproductions for sale by various publishers and distributors. 3rd ed. repr. with suppl. - Washington, D.C.: Library of Congress, 1971. 20 p. and 8 p.

(181) Maps of the American Revolution, a preliminary survey. In: *Quarterly Journal of the Library of Congress*, 28, 1971, p.196-215.
Repr. in: *Canadian Cartographer*, 10, 1973, p.1-20, ill.

(182) Map collections in the Soviet Union and the Democratic Republic of Germany. In: SLA Geography and Map Division *Bulletin*, no. 84, 1971, p.20-23, 39.

(183) (With David K. Carrington) Machine-Readable Map Cataloging in the Library of Congress. In: *Special Libraries*, 62, 1971, p.343-352.

(184) Philip Lee Phillips, Cartobibliographer. In: Karten in Bibliotheken, Festgabe für Heinrich Kramm zur Vollendung seines 65. Lebensjahres. Dargebracht von Lothar Zögner. Bonn-Bad Godesberg 1971, p.95-109. (Kartensammlung und Kartendokumentation. 9.)
Repr. in: *Surveying and Mapping*, 32, 1972, p.349-361.

1972

(185) À la carte. Selected papers on maps and atlases. Comp. by Walter W. Ristow. - Washington, D.C.: Library of Congress, 1972. X, 232 p., ill.
Includes 20 papers, all but two originally published in the Quarterly Journal of the Library of Congress. Ristow is the author of five papers and co-author, and/or editor, of five others, see nos. (186-195).
Repr. 1975.

(186) A manuscript atlas by Battista Agnese. By Lawrence Martin. Ed. by Walter W. Ristow. In: À la carte. - Washington, D.C.: Library of Congress 1972, p.34-38. Originally published in: Library of Congress. Quarterly Journal, 1, 1944.

(187) (With Clara E. LeGear) Sixteenth century atlases presented by Melville Eastham. In: A la carte. - Washington, D.C.: Library of Congress 1972, p.51-61. Repr. and rev. from (81).

(188) America and Africa: Two Seventeenth Century Wall Maps. In: À la carte. - Washington, D.C.: Library of Congress 1972, p.62-75, ill.
Repr. and rev. from (151).

(189) Captain John Smith's Map of Virginia. In: À la carte. - Washington, D.C.: Library of Congress 1972, p.91-95. Repr. from leaflet accompanying a Library of Congress facsimile of John Smith's Map of Virginia. Washington, D.C. 1957.

(190) Augustine Herrman's Map of Virginia and Maryland. In: À la carte. - Washington, D.C.: Library of Congress 1972, p.96-101.
Repr. and rev. from (102).

(191) John Mitchell's Map of the British and French Dominions in North America. Comp. and ed. from various published works of Lawrence Martin by Walter W. Ristow. In: À la carte. - Washington, D.C.: Library of Congress 1972, p.102-113.

(192) From an Actual Survey, Early Maps of Pennsylvania and Virginia. In: À la carte. - Washington, D.C.: Library of Congress 1972, p.152-161.
Also in (146).

(193) John Melish and his Map of the United States. In: À la carte. - Washington, D.C.: Library of Congress 1972, p.162-182.
Repr. and rev. from (116).

(194) South American Historical Maps. By Lawrence Martin. Revised and expanded by Walter W. Ristow. In: À la carte. - Washington, D.C.: Library of Congress 1972, p.189-203.
Repr., with revisions, from Library of Congress. Quarterly Journal, 1, 1944.

(195) John Disturnell's Map of the United Mexican States. By Lawrence Martin..Edited and abridged by Walter W. Ristow. In: À la carte. - Washington, D.C.: Library of Congress 1972, p.204-221.
Repr., with revisions, from Treaties and Other International Acts of the United States of America, vol. 5, 1937.

(196) [Review of] DeBrahm's Report of the General Survey in the Southern District of North America. Ed. by Louis De Vorsey, Jr. Columbia, S.C. 1971. In: *Surveying and Mapping*, 32, 1972, p.271-272.

(197) The anastatic process in map reproduction. In: *Cartographic Journal*, 1972, p.37-42, ill.

(198) Captain John Smith's Map of Virginia. [Text to accompany facsimile edition of the map of 1624.] Rev. and repr. ed. - Washington, D.C.: Library of Congress 1972. 4 p. (Library of Congress Facsimile no. 12.)

(199) Ena L. Yonge 1895-1971. [Obituary] In: SLA Geography and Map Division. *Bulletin*, 88, 1972, p.24-29, portr.

(200) [Foreword in] Guthorn, Peter J.: British maps of the American Revolution. - Monmouth Beach, N.J.: Philip Freneau Press 1972 [i.e. 1973].

1973

(201) Guide to the History of Cartography. An annotated list of references on the history of maps and mapmaking. - Washington, D.C.: Library of Congress 1973. 96 p.

(202) The Bibliography of Cartography. - Boston: G.K. Hall 1973. Introduction by Walter W. Ristow, p.III-V.

(203) [Foreword to] A Descriptive List of Treasure Maps and Charts in the Library of Congress. 2nd ed. Comp. by Donald A. Wise. Washington, D.C.: Library of Congress 1973.

1974

(204) Rand McNally Road Atlas, United States, Canada, Mexico. Golden Anniversary Edition. Chicago, New York 1974. - Introduction; Consultant for roads, road maps.

(205) [Review of] Howse, Derek; Michael Sanderson: The Sea Chart. New York 1973. In: *American Cartographer*, 1, 1974, no. 1, p.88.

(206) Map production and procurement today. In: SLA Geography and Map Division. *Bulletin*, no. 96, 1974, p.2-9, 17.

(207) Dutch polder maps. In: *Quarterly Journal of the Library of Congress*, 31, 1974, p.136-149, ill.

(208) Cartographic information services of the Library of Congress. In: *American Cartographer*, 1, 1974, p.125-130.

(209) Introduction to facsimile edition of Thomas Jefferys' American Atlas (London, 1776). - Amsterdam: Theatrum Orbis Terrarum 1974, p.V-XI.

(210) Geography and Map Libraries Subsection, 40th General Council, International Federation of Library Associations. In: SLA Geography and Map Division. *Bulletin*, no. 98, 1974, p.69-71.

1975

(211) Cartographic information services of the Library of Congress. In: Puti Razvitija Kartografii [The developing course of cartography]. Festschrift honoring K.A. Salichtchev on his 70th birthday. Moskva: Izd. Moskovskogo Universiteta 1975, p.222-228.

(212) Lithography and maps. In: Five Centuries of Map Printing. Ed. by David Woodward. Chicago: University of Chicago Press 1975, p.77-112.

(213) The Geography and Map Division, a Guide to its Collections and Services. - Washington, D.C.: Library of Congress 1975, 42 p., ill.

(214) The emergence of maps in libraries. In: Map Librarianship: Readings. Comp. by Roman Drazniowsky. Metuchen, N.J.: Scarecrow Press 1975, p.451-489.
Repr. from (41).

(215) Library of Congress Geography and Map Division Map Processing Project. In: SLA Geography and Map Division. *Bulletin*, 102, 1975, p.30-33.

Also in: Library of Congress. *Information Bulletin*, 34, 1975, no. 34, Appendix I, p.A-198 - A-199.

(216) Conference on History of Cartography held (Greenwich, Sept. 7-11, 1975). In: SLA Geography and Map Division. *Bulletin*, 102, 1975, p.23.

(217) Map Cataloging and Classification in the Library of Congress. - Washington, D.C. 1975. 18 p. Multilith.

1976

(218) World Directory of Map Collections. Comp. by the Geography and Map Libraries Sub-Section [International Federation of Library Associations]. Ed. by Walter W. Ristow. - München: Verlag Dokumentation 1976. 326 p. (IFLA Publications. 8.)

(219) [Foreword to] Union List of Sanborn Fire Insurance Maps held by Institutions in the United States and Canada. Vol. 1 (Alabama to Missouri), by R. Philip Hoehn. [Santa Cruz, Cal.]: Western Association of Map Libraries 1976. (Occasional Paper no. 2.) Foreword p.V-VII.

(220) The Hauslab-Liechtenstein Map Collection. A preliminary Sampling. In: American Congress on Surveying and Mapping. *Proceedings*, 36, 1976, p.205-211.

(221) (Anonym) Anastatic process. In: International Cartographic Association. Map-making to 1900, an historical glossary of cartographic innovations and their diffusion. Ed. by Helen Wallis, Moscow 1976, p.42-43.

(222) Geography and Map Libraries Subsection, International Federation of Library Associations. In: Special Libraries Association. Geography and Map Division. *Bulletin*, no. 103, 1976, p.29-32.

(223) Map library education in the United States and Canada. In: Special Libraries Association. Geography and Map Division. *Bulletin*, no. 104, 1976, p.38-41. Also in: *INSPEL*, 11, 1976, no. 3, p.95-99. Paper presented at the meeting of the Geography and Map Libraries Subsection, IFLA, Oslo, August 1975.

1977

(224) Robert Mills' Atlas of South Carolina. In: *Quarterly Journal of the Library of Congress*, 34, 1977, p.52-66.

(225) Maps for an emerging nation: Commercial cartography in nineteenth-century America. [Catalog of] an exhibition at the Library of Congress. Washington, D.C. 1977. 66 p., ill.

(226) (With R.A. Skelton) Nautical charts on vellum in the Library of Congress. - Washington, D.C.: Library of Congress 1977. 44 p., ill.

(227) The greening of map librarianship. In: SLA. Geography and Map Division. *Bulletin*, no. 111, 1978, p.2-9. [Rev. paper pres. at a meeting of the SLA Geography and Map Division, New York, June 1977.]

Editorial work

Acta Cartographica. A series of monographs and studies on the history of cartography, reprinted from periodicals since 1800. - Amsterdam: Theatrum Orbis Terrarum 1967-
Advisory Editor from 1967-

Imago Mundi. The Journal of the International Society for the History of Cartography. (1937-1974: A review of early cartography.)
Corresponding Editor from vol. 21, 1967-

The Canadian Cartographer. Editor: B.V. Gutsell. - Toronto: York University 1963-
Contributing Editor from vol. 5, 1968-

Surveying and Mapping. Official Quarterly Journal of the American Congress on Surveying and Mapping. - Washington 1940-
Compiler of the regular feature "Distinctive recent maps", vol. 8, 1948 - vol. 28, 1968.

A BRIEF HISTORY OF THE LIBRARY OF CONGRESS GEOGRAPHY AND MAP DIVISION, 1897-1978

By John A. Wolter
Andrew M. Modelski
Richard W. Stephenson
David K. Carrington

Prologue 1800 - 1896

Maps, charts and atlases were included in the collections of the Library of Congress at its establishment in 1800, the year the Federal Government was relocated from Philadelphia to the new city of Washington. It was not until 97 years later, however, that a separate division was created to administer the map collection. This event coincided with the opening of the Library's new building and the appointment of Philip Lee Phillips as the first chief of the new Hall of Maps and Charts.

On April 22, 1800, President John Adams signed an appropriation act for moving the government to Washington. The fifth section of the act, approved on April 24, allocated funds "for the purchase of such books as may be necessary for the use of Congress in the said city of Washington, and for fitting up a suitable apartment for containing them". This section also approved the creation of a joint committee of Congress to prepare a list of needed books and to establish regulations for the use of the Library. A list of desired items was prepared, and in June an order was placed with the London dealers of Cadell and Davies for the needed items. At year's end, the English dealers forwarded to the United States eleven trunks containing 740 books and a case of maps.

The invoice to the shipment indicates that the special map case cost over 5 shillings, and included Arrowsmith's two maps of America, and Faden's map of South America, as well as two non-cartographic wall charts of chronology and biography.Each of the items was described in the invoice as being mounted on canvas and rollers, as was the fashion of the day.

Included in the shipment of books were volumes of maps accompanying such popular works of the day as William Guthrie's *A New System of Modern Geography*, and Clement Cruttwell's *The New Universal Gazetteer*. The shipping invoice also notes the inclusion of a separate book of maps entitled the *American Atlas*, a title common to several of the popular atlases of the day. The 1812 printed catalog of the Library clarifies the issue by noting that the *American Atlas* belonging to Congress was, in fact, an 1800 edition of the famous atlas of this title compiled by Thomas Jefferys, Geographer to the King of England, and published posthumously beginning in 1771 by the London firm of Sayer and Bennett. Jefferys' was truly one of the best atlases of America published in the eighteenth century. It presented, in one volume, the works of such outstanding men as Captain James Cook, the mariner; Lewis Evans, American geographer and scientist; Samuel Dunn, English mathematician; Major Samuel Holland, Surveyor General of the Northern District in America; Jean D'Anville, French geographer; Thomas Pownall, Governor of Massachusetts; and William Scull, Joshua Fry, Peter Jefferson, and Henry Mouzon, American colonial surveyors.

The books and maps ordered from Cadell and Davies arrived in Baltimore in February 1801, aboard the ship *American*, were transshipped via packet to the port of Georgetown, and thence by wagon to the office of the Clerk of the Senate. Here they remained unopened until Congress convened in December 1801. A joint committee was formed respecting books and maps purchased pursuant to a late act of Congress, and to make a report for the future arrangement of the same. John Randolph of Virginia drafted the committee report and on January 26, 1802, an act concerning the Library for the use of both houses of Congress was approved. The new legislation directed that the recently purchased books and maps, together with the books "which have heretofore been kept separately by each House, shall be placed in the Capitol, in the room which was occupied by the House of Representatives, during the last session of the Sixth Congress". Apparently members realized very early the perishable nature of maps and stated in section four of the new law that "no map shall be permitted to be taken out of the Library by any person".

The books and maps were sorted and arranged by April 1802, and the first printed catalog of the Library of Congress was issued. This brief 10-page work, entitled *Catalogue of Books, Maps, and Charts, belonging to the Library of the Two Houses of Congress* (Washington, Printed by William Duane, April 1802), is an invaluable source of information about the early collections of the Library.

A decade later the Library possessed some 50 maps and an unrecorded number of atlases. These and most of the other holdings of the Library were destroyed in 1814, when British soldiers burned the Capitol.

To reconstitute the Library, Thomas Jefferson offered his personal collection of books and other reference materials, which he had carefully and selectively assembled over a period of 50 years. Fortunately, the Congress approved the purchase of this private scholarly library, and the materials were transported from Monticello to Washington in horse-drawn wagons in the spring of 1815. Geography was one of the intellectual interests of the well-read Jefferson, and a number of general and regional geographical books, as well as a rich selection of atlases and maps, were included in the acquisition. Many of these items are described in volume four of the five-volume *Catalogue of the Library of Thomas Jefferson*, which was compiled, with annotations, by E. Millicent Sowerby and published by the Library of Congress (1952-59).

During the next several decades additional cartographical items were acquired by purchase, gift, exchange, and official deposit.

Evidence of Congressional interest in maps appears from time to time in votes of the Joint Committee on the Library and in requests to the Librarian for procuring specific items. On March 29, 1830, for example, the Committee requested that the Librarian be instructed to acquire David H. Burr's 1829 county atlas of the State of New York, and the best maps of the several states which were not already in the Library.

There is no record of the size of the map and atlas collections on December 24, 1851 when fire again swept through the Library's rooms in the Capitol. The official report on the conflagration, issued on January 7, 1852, indicated that some 35,000 books were destroyed as well as "nearly all our extensive collection of Maps". The collections were once again reconstituted, but without the nucleus of a Jefferson library. Maps and atlases, published in greater numbers after the middle of the 19th century, were among the routine acquisitions.

During the 1850's, three unofficial proposals to establish a separate map department in the Library were advanced. The first was made by Lt. Edward B. Hunt, U.S. Army, at the annual meeting of the American Association for the Advancement of Science, held in Cleveland from July 28 to August 2, 1853. Lieutenant Hunt, who was then on detail to the U.S. Coast Survey, complained that "no [map] collection exists in our land which furnishes full materials for extensive investigations, such as are now more and more demanded by questions of history, science, commerce, and policy". He proposed, therefore, that there be established, preferably at the Library of Congress, "a complete and special geographical library, ... [which] would be a valuable aid to all ... researchers". Because of the difficulties he experienced in conducting his personal research, Lieutenant Hunt further stressed the need "to arrange complete and systematic indexes and catalogs, which would at once make known all the materials so arranged as best to facilitate special research".

Dr. Johann Georg Kohl reiterated two years later the need for a systematic cartographic collection. He proposed the establishment of a national cartographic depot in the Library of Congress in a lecture at the Smithsonian Institution which was published in its *Annual Report* for 1856. Kohl, an eminent German geographer, historian, librarian, Americanist, and world traveler had, during the years of his travel and study, painstakingly hand-copied, from original sources in the largest libraries and museums of Europe, numerous very early maps showing the gradual development of knowledge of North America. In 1854 he brought this valuable collection of maps to the United States.

At that time, these maps represented the only comprehensive collection of hand-copied or facsimile maps which existed in America derived from original sources in the leading libraries and museums of Europe. Their existence filled a gap in available reference information for American geographers, historians, and cartographers of the late 19th century.

The U.S. Congress made a grant to Kohl in 1856 to prepare hand copies of maps from his European collection for consultation by American scholars. It served them well until the use of photoreproduction for library material became established in the early 20th century. Kohl's and his American assistants' hand copies of maps prepared as a result of the grant were originally secured for the State Department by William Learned Marcy, Secretary of State. During the Civil War, the entire collection came temporarily under the custody of the War Department, was stored in a troop billet, and came very close to being destroyed. On July 17, 1903, they were transferred from the Bureau of Rolls and Library, Department of State, to the Division of Maps and Charts, Library of Congress, through the efforts of Philip Lee Phillips.

Prof. Daniel Coit Gilman of Yale University included a plea for a federal map collection in an article entitled "The Last Ten Years of Geographical Work in this Country", published in the *Journal of the American Geographical Society* (vol. 3, 1872, p.111-113). Gilman acknowledged that "many able officers of the government [were] engaged in investigations of the highest value to mankind, [but] much of their usefulness is impaired by the defective arrangements for gathering up, presenting and distributing to the public the results thus ascertained. They likewise point to the importance," Gilman continued, "of having established in Washington, or elsewhere, as a department of the general government, a bureau of maps and charts and geographical memoirs, where all these vast accumulations may be stored, classified and rendered accessible, like the books in the Library of Congress ... so that persons who have the right may make inquiry respecting them."

The successive recommendations of Hunt, Kohl, and Gilman were not acted upon, and several more decades passed before a separate map department was established in the Library of Congress. Nonetheless, during these years the cartographic collections of the Library continued to be augmented through various acquisitions. The volume of accessions, in all formats, increased greatly following passage of the Copyright Act in 1870, which required mandatory deposit in the Library of Congress of two copies of all copyrighted publications.

Several noteworthy collections of maps were acquired by purchase before 1897. The earliest such accession was the Faden Collection which includes 101 manuscript and printed maps relating to the French and Indian Wars and the American Revolution. The maps, many of which were drawn by British military engineers, had formed part of the working collection of William Faden, foremost English map publisher during the last quarter of the 18th century, and Geographer to King George III of England. It was acquired for the Library of Congress, in June 1864, by the Joint Library Committee of Congress from Edward Everett Hale. One thousand dollars were appropriated by Congress to cover the purchase cost. Nathan Hale (nephew of the Revolutionary hero of the same name), father of Edward Everett, purchased the collection from a Reverend Mr. Converse, who in turn had obtained it at the sale of Faden's estate.

In 1867 the Congress purchased for the Library, at a cost of $100,000, the notable collection which had been assembled by Peter Force, distinguished archivist and Americanist. In addition to extensive and valuable holdings of books, pamphlets, manuscripts, journals, and newspapers, the Force purchase included more than 1,200 maps and views. The maps greatly enriched the Library's cartographic holdings. After more than a century of collecting and growth, items included in the Force purchase are still among the rarest treasures in the map collections. Peter Force's library is considered to be one of the finest collections of Americana ever assembled by one man.

In May 1882 the Library acquired at an auction in New York City a collection of 20 colored manuscript maps of New Jersey prepared by John Hills, a British army engineer on the staff of Sir Henry Clinton, British Supreme Commander in America from 1778 to 1781. Until earlier in the same year, the collection had been among Clinton's papers. The papers, letters, journals, and maps belonging to the Comte de Rochambeau, the Commander-in-Chief of the French Army in America during the Revolutionary War, were purchased for the Library in 1883, from the Marquis de Rochambeau. The collection of maps included in this purchase consists of 40 manuscript maps, 28 printed maps, and a manuscript atlas.

Before the establishment of federal mapping agencies, maps prepared under official authorization were included as supplements in published congressional document series. The maps were printed on contract by various engravers and lithographers. In 1843 some 150 congressional series maps were assembled and bound in a large folio volume, now in the Geography and Map Division. Although an inscription states that "this is not a complete collection of maps, charts etc., published by order of Congress", the volume provides an excellent cross section of United States official mapping over a period of more than two decades.

These and many other special Division collections acquired during the past century and a half are described in the 1975 edition of *The Geography and Map Division; A Guide to its Collections and Services*, Washington, D.C., Library of Congress, 1975. 42 p.

In 1875 Philip Lee Phillips joined the staff of the Library of Congress. His duties included servicing books and general maintenance of the collection, in addition to cataloging. Early in his career he became aware of the disorganized state of the Library's cartographic materials. In time that he could spare from his regular duties, Phillips began to organize and catalog the Library's maps and atlases. Because of these efforts and his growing knowledge of maps, their history, and their makers, Phillips came to be recognized as the cartographic specialist. As early as 1878 he was also at work compiling bibliographic references, not only for his monumental *A List of Maps of America*, but also for a list which eventually grew to be the most comprehensive and unique analytical index to the worldwide literature of cartography. With little hope of obtaining authorization to publish his Bibliography of Cartography separately, he inserted it as a

The Hall of Maps and Charts, October 1898

preface of 90-pages in the *List of Maps of America.* In reporting the publication of the *List...*, to the Librarian of Congress in 1901, Phillips noted that he hoped in time to add to and improve the *Bibliography of Cartography* and make it a work of more extensive scope. Phillips and his staff continued to add entries to the card file, and he continued to recommend it as a publication. However, it was not published until 1973 when G.K. Hall and Company issued the cartobibliography in five volumes from the 100,000 accumulated entries. A supplementary volume of about 25,000 entries will be published by G.K. Hall within the next few months.

The Early Years 1897 - 1947

The Library of Congress was moved during the summer and early fall of 1897 from its crowded quarters in the Capitol Building to its beautiful new building. At that time a Hall of Maps and Charts was organized with Philip Lee Phillips as Superintendent. Mr. Phillips' first report, dated September 22, 1897, to Librarian of Congress John Russell Young, noted the following items of special interest: the collection was estimated at 27,000 maps and 1,200 atlases; he had been allowed two assistants, one of whom was selected; a classification scheme for maps had been developed; and an acquisition program was outlined.

Phillips had unusual skill in being able to interpret requests and to anticipate bibliographies which would be helpful to readers and correspondents. During the first year after the Division was organized, he asked the Librarian for funds to publish the *List of Maps of America*. As previously noted, the *List of Maps of America* was printed in 1901. Between 1909 and 1920 the first four volumes of the *List of Geographical Atlases in the Library of Congress* were compiled and issued as a Library of Congress publication. In addition, the Division was able to prepare several other comprehensive bibliographies, some of which have, unfortunately, never been published.

In 1898 the Army and Navy called upon the Library of Congress for cartographic information for use in prosecuting the Spanish American War. The same service was again requested during

World War I. In 1918, Mr. Phillips reported to the Librarian that the Division has been found adequately equipped to meet the demands of the various departments of the government for maps to be used in the prosecution of the War, and that the officials of the Navy and War Departments had repeatedly assured him that items were secured which were absolutely necessary for immediate use. The completeness of the Division's collection of various European Government staff maps was especially noteworthy.

Again, in the World War II period, 1940-1945, the Division of Maps was used extensively by Federal Government departments. Colonel Lawrence Martin, then chief of the Division, reported in 1943, that the continued impact of the World War upon the Division of Maps was so impressive that some of the summaries of service rendered deserved comparison with earlier records. The service was fifteen times as great as in 1938-39, a normal year previous to the war.

Another important service which was rendered, namely furnishing current and historical cartographic data for use in the study of boundary and other problems often occupied the attention of the first two chiefs, Mr. Phillips and Colonel Martin. Several cases in which the Division rendered valuable aid as noted in various annual reports of the Chief of the Division are included below:

1898	Venezuelan Commission (Venezuelan-British Guiana Boundary Commission)
1903	Honduras and Guatemala boundary question
1922	Boundary dispute between Canada and Newfoundland
1926	Texas-New Mexico boundary case
1928	... study of the question of the boundaries between high seas and territorial waters along the coast of the United States. (At the request of one of the American members of the Committee of Experts of the League of Nations for the Progressive Codification of International Law).
1930	Minnesota-Ontario boundary (Dispute over 1,000,000 acres of Hunter's Island).

1930 Horseshoe Bend on the Mississippi River (Arkansas-Tennessee boundary question)

1930 British Borneo-Philippine Island boundary settlement

1931 New Jersey-Delaware boundary

1934-36 Boundary line between the District of Columbia and the State of Virginia

1936 Boundaries between Michigan, Minnesota and Wisconsin in Lake Superior

Reference Activities

The role that the Hall of Maps and Charts and its successors, the Division of Maps and Charts, the Division of Maps, the Map Division, and the Geography and Map Division was to play in the reference, processing, acquisition, and bibliographic programs of the Library of Congress was foretold by the activities during the early years of the division. After the establishment of the Hall of Maps and Charts, the Library's map and atlas collections immediately became more easily accessible and valuable reference sources for Congress, federal agencies, and departments, as well as for international and foreign commissions studying boundary and other problems, and for serious scholars.

Congressional requests for cartographic materials have always varied from year to year, depending on the problems currently under consideration. Mr. Phillips noted in his 1901 annual report that assistance had been given to various committees of Congress and he mentions especially the District of Columbia and Park Committees.

The information in subsequent annual reports shows the kind of requests initiated by Congress also differed widely. For example, information requested varied from data on the location of some geographical feature to requests for lengthy bibliographies with critical annotations. The record of the first fifty years also indicates that there was no consistent increase or decrease in the use Congress made of the Library's cartographic holdings. Instead, there were periods of active interest, as well as periods when little use was made of atlases, maps or other cartographic materials.

Scholars interested in some phase of cartographic history or wishing map information as an aid in solving research problems soon learned that the Library of Congress had an outstanding collection of maps of the Americas and many other areas. A list of works published by scholars who have used the cartographic resources of the Library of Congress would be long indeed. It might be of interest to mention a few of the scholars who were using the resources of the Division during this early period.

In 1901 Marcus Baker was compiling a list of surveys and maps of the District of Columbia, and Edward M. Mathews was working on a bibliography of the cartography of Maryland. Mr. Phillips remarked in 1916 about the great amount of work necessary in aiding Dr. Charles O. Paullin and Dr. James A. Robertson of the Carnegie Institution in connection with the *Atlas of the Historical Geography of the United States*, published in 1932. The American Geographical Society in 1926 had a staff member in the Division for more than two months working on a list of *Maps of Hispanic America*. Beginning in 1928, Professor L.C. Karpinski was working on a list of maps of Michigan not found in the libraries of Michigan. This work was published by the Michigan Historical Commission in 1931. In 1937, Carl I. Wheat published a list of early maps showing California Gold Fields. He used Division of Maps resources and staff extensively over a period of many years for that work and many others culminating in his monumental six volume *Mapping the Transmississippi West, 1815-1861*.

Colonel Lawrence Martin, glaciologist and geographer succeeded Mr. Phillips as Chief of the Division in 1924 and served in that capacity until 1944. He retired from government service in 1946. Although not as prolific an author and bibliographer as Phillips, his contributions were many and as previously noted, several important boundary disputes were also helped to settlement by his service as a consultant.

Unlike his predeccessor Phillips, Martin was a member and officer of several professional organizations, including the

Association of American Geographers and the Geological Society of America. He also served on the U.S. Geographic Board, the U.S. Board of Surveys and Maps, the U.S. Board on Geographic Names and the National Research Council, organizations which his successors over the past several decades have also served in various capacities.

In addition to his official publications which are included in the list at the end of this paper, Colonel Martin prepared many scholarly and interpretive articles which were included in publications of other government departments, or were issued as articles in scientific or professional journals.

Processing Activities

Phillips' main interest was in gaining bibliographic control of all items in the custody of the Division of Maps and Charts. As has been seen, his first step toward this goal was actually started twenty years prior to his appointment as Superintendent of the Hall of Maps and Charts when he began compilation of his noteworthy *List of Maps of America*. He had been in charge of the cartographic collections for less than a month when he submitted a classification scheme for maps. Also, he devised the method of placing each map, whether it was one or several sheets, within a folder and "titling" the folder with information regarding area, subject, date, scale, and author, in that order. This classification scheme has made the maps in the Library of Congress fairly easy to locate in the files for many years. It is also the forerunner of the atlas and map sections of the Library of Congress Classification Class G.

Other processing activities developed by Phillips which have proved their value in succeeding years include the practice of cataloging all atlases and including tables of contents for many (this activity has resulted in the eight volume *List of Geographical Atlases* described in the appended publications list); the compilation of a catalog of copyrighted maps by clipping copyright catalogs and pasting entries on cards which were then arranged in card catalog form;

Philip Lee Phillips, first chief of the Division, 1897-1924

the compilation of checklists of special material in constant use (such as a list of maps and views of Washington, D.C. and Mount Vernon, Boston, etc.), some of which were published.

Col. Martin, in general, continued the processing methods initiated by Phillips. He did, however, simplify the procedure for cataloging atlases and had a card catalog prepared for the general maps of the United States from 1776 to date.

The greatest problem faced by every Division administrator was how to gain control of the vast quantity of individual maps. Phillips initiated several projects designed to catalog these maps, as did his successors, Colonel Martin, Dr. Platt and Dr. Adkinson. None was able to produce a workable solution. Adkinson noted that bibliographic control of separate maps was a problem that had to be solved before the Disivion of Maps could effectively and economically give the quality of reference service desired. As will subsequently be seen, it was not until 1968 and the advent of MARC Map that this control became both feasible and possible.

The Library of Congress map and atlas collection slowly won recognition as one of the most comprehensive in the world. This recognition, of course, validates the acquisition policies of all the former chiefs more than a detailed analysis of items can ever do.

Philipps, in his first year earnestly recommended that early maps of America should be obtained. He noted that it was possible to strengthen the collection by exchanges with other libraries, and that there should be relief maps in plaster of America, Europe, Asia, and Africa, with large globes displaying the latest achievements in geographical research. He stated that the list of county and city atlases should be completed, and an effort should be made to obtain through the aid of consuls, the latest maps of South America and Europe.

In the first two decades of the 20th century, Phillips made trips to Europe, during which he contacted libraries, official map producing agencies and private dealers.

International exchange of maps has always been difficult to arrange, but many were begun by Phillips quite early, and special efforts have been made at various times to increase this form of exchange. In every case this has been of mutual benefit to those concerned.

The practice of depositing in the Division two copies of all maps published by federal agencies makes the Library's cartographic collections very strong in United States materials. This procedure was outlined by Phillips and he succeeded in getting federal agencies to follow the practice at an early date. His particular interest in atlases, maps pertaining to the Americas, and maps of local areas of the United States is shown by the present strength of collections in these three fields. In 1902, Mr. Phillips reports writing to officials in 2,857 U.S. counties requesting maps and atlases. This program was an apparent success in 1902, as the Division received some 4,635 maps and 261 atlases. This procedure was repeated in 1915 and 1921 and at stated intervals since that time.

During this early period, too, the number of items received as copyright deposits was of great significance, particularly what was known early on as the Sanborn-Parris Fire Insurance Map collection.

Both Mr. Phillips and Colonel Martin spent much time checking dealers' catalogs, as well as developing exchanges with other domestic libraries. By these and other methods, they developed a strong collection of maps and atlases dealing with the early history of the Americas and more particularly of the United States.

The Division continually worked toward the acquisition of all editions and impressions of maps significant in boundary studies and in strengthening the Division's foreign map holdings; especially the large scale official maps of European countries. This policy was of enormous benefit to the Federal Government during World War II.

Phillips complained as early as 1919, however, that foreign governments were not very generous with their cartographic publications. Since the active operations of World War I did

Map Reading Room, ca. 1927

not cover a wide geographical area, there was little need for the extensive map production that took place during World War II. Accessions figures for 1945 to 1947 show the results of the world wide operations of the latter war and the consequent mass production of maps.

In summarizing the acquisitions program of the Division during the first 50 years, attention is called to TABLE 1.

S t a f f

The size and quality of the staff is most important in the effective functioning of any organization. During most of the existence of the Geography and Map Division, there has been a large proportion of the staff who were acquainted with the arrangement of the collection and familiar with the significance and content of a considerable portion of the material contained therein. Except for the World War I and II years, long periods of service by employees have enabled the Division to maintain an experienced and well-trained staff.

Mr. Phillips began with a staff of two persons and a map mounter in the Hall of Maps and Charts. In 1939 there were 9 members on the staff of the Division of Maps, and by 1946 the staff had increased to 16 persons. During the same period the collection had grown from less than 30,000 items to more than two million items. With the increasing ratio of items to staff member there was a concomittant decrease in the amount of processing and bibliographic research done in the Division. To illustrate, in 1901 there were less than 10,000 items per staff member and the Division was actively engaged in compiling bibliographies and in cataloging atlases and American maps. By 1909 Phillips was complaining about the lack of staff and indicated in his report that the bibliographic research of the Division was suffering. His staff of six had to service a collection of one-half million items.

During much of its early existence, the Division had to work under the handicap of too few persons in crowded and unsatisfactory work spaces for processing activities. Those problems became increasingly troublesome each year. In 1947 a staff of

16 was servicing a collection of more than two million items or approximately 125,000 items per staff member. To add to the above load, the number of items placed before readers had increased from fewer than 10,000 per year in 1905 to more than 40,000 items during the three year period 1944-1947.

Summary

A summary of the first 50 years shows that during the period 1897 to 1924, the major interest of the chief was the publication of checklists and bibliographies. The energies of the entire Division were centered on this program, while reference services and acquisitions of materials seemed to be of lesser importance. Of special interest to Phillips were bibliographies of maps and atlases pertaining to the Americas or parts thereof. His acquisitions program was oriented toward the acquisition of such materials and the Library's collections soon became strong in these areas.

Such foreign maps and atlases as Mr. Phillips recommended for procurement came as a result of international exchanges or were selected to give the Library a representative collection of maps of foreign areas considered of importance to the United States.

Mr. Phillips, for whatever reason, did not seem interested in working actively with groups of geographers or cartographers. This is borne out by the fact that he did not attend meetings of the Board of Surveys and Maps of the United States Geographic Board, nor did he participate in professional meetings of geographers or cartographers.

Col. Lawrence Martin changed the emphasis of the Division's activities soon after he became chief in 1924. The active program of publishing checklists and bibliographies slowed. Emphasis was placed on interpreting materials in the collection through personal and correspondence reference service, by active participation in professional meetings of geographers and geologists, and by preparing scholarly analyses of individual maps and of boundary problems.

Between 1942 and 1945 the staff was fully occupied with servicing government agencies with maps covering areas of strategic and operational importance. From 1945 through 1947 emphasis was on handling the flood of maps published for military and naval planning and operations and in serving the Federal government agencies and the Congress with cartographic materials needed for planning and other post-war activities. As shall be subsequently seen, this flood of maps had many beneficial effects for the Division and its users.

Transition 1947 - 1950

Colonel Martin served as chief of the Division, except for a brief period during World War II when he served in the Office of Strategic Services, until June 1946 when he retired from government service. At that time Librarian of Congress Luther H. Evans appointed him Honorary Consultant in Geography and he served the Library in this capacity for several years. Colonel Martin died on February 12, 1955, after a long illness.

While Martin was completing his OSS duties, Dr. Robert S. Platt, a geographer on leave from the University of Chicago was appointed acting chief of the Division. His tenure was brief, and ended in May 1946 when he returned to his teaching post. However, he made recommendations for a restructuring of the Division along the lines of a geographical research institute. Some of them were incorporated in Dr. Burton W. Adkinson's general reorganization plan instituted in 1948. Adkinson, who came to the Division in mid-1945 as assistant chief, became chief in October 1947, and Director of the Library's Reference Department two years later in November 1949. He left the Library of Congress in November 1957 when appointed Head, Office of Scientific Information, National Science Foundation.

Dr. Walter W. Ristow was appointed assistant chief of the Division in December 1946, served as acting chief for a period after Adkinson was appointed to the Reference Department and resumed his position upon the appointment of Dr. Arch C. Gerlach as chief in July 1950.

Maturity 1951 - 1978

Until the late 1940's, the Division was not departmentalized. Each employee was expected to spend a portion of his or her time on acquisitions, cataloging, and public services. The three professional librarians below the level of Division chief, worked every third week in the reading room assisting patrons with their requests for maps and atlases. When not assigned to the reading room, they answered reference correspondence, and carried on the other necessary functions of the Division.

Reference Activities

With the expansion of the staff, made necessary after World War II by the need to process the rapidly accumulating backlog of maps produced or captured by our Armed Forces, the Division was reorganized into four administrative units: the Office of the Chief, the Acquisitions Section, the Processing Section, and the Reference Section.

In 1950, when Arch C. Gerlach became division chief, the reference staff consisted of one professional reference librarian and one clerk. The former was responsible for assisting map readers with their needs, answering telephone calls, and drafting responses to letters received by the Library requesting information about maps and atlases. The clerk handled loan and photoduplication requests and assisted the reference librarian in filling the needs of the readers. Even at this early date, the Reference Section was undermanned and it was necessary to call upon staff from the rest of the Division to assist with public services.

With the establishment of a separate Reference Section, however, new emphasis began to be placed on service to the public. For example, a two week sampling was conducted in 1950, to determine what portions of the collections were most used. The sampling revealed "that approximately two-thirds of the reader requests are for maps or information relating to the United States as a whole, ranking second in popularity to maps of the District of Columbia in this group. Maps of individual states and cities are also called for frequently.

The importance of cartographic exhibits as an information and educational tool was reemphasized with seven map exhibits being planned and arranged during the same year. Furthermore, after a lapse several years, monthly map exhibits on current themes were arranged in the Senate Wing of the United States Capitol Building.

In an attempt to rectify the understaffing of the Reference Section, in fiscal year 1952, an atlas cataloger and a map titler were detailed to two newly created reference librarian positions. The staff remained at this size until the position of clerk was lost in January 1953, due to retirement. Increasing demand for maps and cartographic information accompanied by the reduction of the reading room staff to three, required the Division chief, assistant chief, and the bibliographer to be frequently called upon to render assistance. Subsequently, a fourth and fifth reference librarian were added, the clerk's position was regained in fiscal year 1968, and for a short time during this same year the staff was augmented by a sixth librarian.

In 1960, the reference and bibliographical functions of the Division were combined under the direction of Clara Egli LeGear, the Division's distinguished bibliographer. Mrs. LeGear remained as head of the Reference and Bibliography Section until her retirement after 47 years of service on December 15, 1961. The new head of the section was Dorothy W. Bartlett, who at the time of her appointment was serving as head of reference in another Federal map library. Mrs. Bartlett administered the Section until she transferred in February 1970 to the re-instituted position of Bibliographer.

In her new position, Mrs. Bartlett was chiefly responsible for the preparation of new entries for the Division's Bibliography of Cartography which she had been previously doing on a part-time basis. Richard W. Stephenson, who had served as a reference librarian in the Section from 1954 to 1966 and as head of the Division's Acquisitions Section since 1966, was appointed head of the Reference and Bibliography Section in April 1970.

After 55 years in the Main Building of the Library of Congress, the Division was moved to the Annex Building (now the Thomas Jefferson Building. See Table II). A new reading room was established on the ground floor in the Southwest curtain of the building. The Map Reading Room was completely redesigned, and new equipment was purchased. This included eight birch readers' tables, sixteen reader's chairs, desks for the reference librarians, a large magazine rack for displaying current geographic and cartographic periodicals and steel entrance counter with roller shelves for folio atlases assigned to the reading room reference collections. In addition, wallshelving was constructed to house the Section's reference book collection. All equipment was color coordinated to match the birch reader's tables. Thus, the map reading room was now a far more pleasant and effective working area.

As part of a general reallocation of space in 1962, the Map Reading Room was moved from the Southwest to the Northwest curtain, adjacent to the Division's administrative offices. Here it remained until September 1969, when the entire Geography and Map Division moved to temporary rental facilities at 845 South Pickett Street, Alexandria, Virginia.

The one-story, warehouse-type building presently occupied by the Geography and Map Division is situated in the western part of Alexandria, approximately 12 miles from Capitol Hill. Centrally located in the building, which is officially designated as the Pickett Street Annex, is an attractive, carpeted reading room containing some 3,000 square feet of floor space. Here, members of Congress, federal employees, and the general public may consult maps, atlases and related publications in comfortable, well-furnished modern surroundings, capable of accomodating 16 readers. The Reading Room is open to the public Monday through Friday from 8:30 a.m. to 5:00 p.m. and on Saturdays from 8:30 a.m. to 12:30 p.m. Furnished as before, the Reading Room is equipped with a visual display computer terminal and printer which provides access to several of the Library of Congress' automated data bases. Also included in the reading room is a light table for tracing maps and a microform reader which is capable of handling both 35 mm microfilm and microfiche.

Geography and Map Reading Room, April 1972

A reference book collection containing some 8,000 volumes, chiefly in the subject fields of bibliography, cartography, geography, history, and toponymy is maintained and current editions of some 200 geographic and cartographic journals are displayed on a periodical rack in the reading room. For the most part, the division keeps only the most recent editions of the journals. Bound volumes of important professional journals, however, are shelved in the Division's reference book collection.

Rare cartographic materials are consulted in an enclosure seating four persons immediately adjoining the rarity vault. In wooden card file drawers surrounding the enclosure are the Division's dictionary atlas catalog containing citations to a major segment of the atlas collection, and the MARC Map catalog. The latter, the fastest growing bibliographic record in the Division, includes author, subject, title, and series citations for over 55,000 general and thematic maps received or recataloged by the Division since 1968.

A specially designed vault, with temperature and humidity controls separate from the rest of the building is adjacent to the reading room. It contains some 5,000 square feet of floor space and provides excellent fire and security protection for some of the nation's rarest cartographic treasures.

Use of the Collections

Initially, the move of the Geography and Map Division from Capitol Hill had a serious negative effect on the public services. No longer was it convenient for library patrons (particularly those without cars) and Library of Congress staff members to consult maps, atlases, and related materials As anticipated, there was an appreciable increase in telephone inquiries to the Reading Room. Telephone calls rose by almost 29 percent with calls from other units of the Library of Congress up by 71 percent.

In the nearly nine years since the move to Picket Street Annex, public services have steadily increased until in fiscal year 1977, when total direct reference services performed by the Division actually exceeded the level reached before the move from Capitol Hill. Reference statistics for the period October 1976 to September 1977 indicate that 3,248 readers registered in the map reading room, and they were assisted 8,251 times. In addition, 5,234 telephone inquiries were received and 3,857 letters and memos were prepared. Total direct reference services, therefore, totaled 17,342. See Table III for comparative use statistics.

E x h i b i t s

Traditionally the Division has encouraged and supported a strong exhibition program. Writing in 1954, Walter Ristow noted that "people are interested in maps and in the information they portray. As map librarians and geographers, it is our pleasure and responsibility to stimulate and broaden that interest". Cartographic exhibits in the Library of Congress over the years have focused on anniversaries and holidays, newsworthy events and places in the news, historic occasions, and current acquisitions.

Prior to its move from the first floor of the main building of the Library of Congress in 1953, the Geography and Map Division had an ample display area along the entrance hall to the Map Reading Room. In addition, the Division was responsible for preparing current events map displays in the lobby of the United States Senate and for maintaining an exhibit case at the west basement entrance to the Library of Congress in which were featured the "News Map of the Week" and other maps of current interest.

The large display area at the entrance to the Reading Room permitted Division personnel to mount regular cartographic exhibits on many themes. In 1949, for example, representative examples of the well-known map supplements issued by the National Geographic Society since 1899 were displayed. The exhibit recognized Dr. Gilbert Grosvenor's half century association with the National Geographic Society. In May of the following year, an exhibition was held which showed eighteenth century plans of Old World cities and their influence on

L'Enfant's 1791 plan of Washington, D.C., and later the same year a display was mounted with commemorated the 300th anniversary of the birth of Fr. Vincenzo Coronelli, the eminent Italian cartographer. The latter exhibit featured the Library's excellent set of gores for Coronelli's large terrestrial globe of 1688.

In 1953, in order to obtain much needed space for its burgeoning collections and to centralize its operations, the Division moved to the Annex Building (now called the Thomas Jefferson Building) situated to the rear of the Library's main building. The crucial need for office and collection space forced the Division to sacrifice the magnificent exhibit hall available in its old location. Cartographic exhibits were greatly curtailed following the move. In 1959, an exhibits board was installed in the Map Reading Room and regularly, thereafter, small cartographic exhibits were mounted. Although by necessity, major cartographic exhibits were limited during the remainder of the 1950s and the 1960s, maps from the collections were regularly used to illustrate the Library's general exhibitions. Especially noteworthy during this period were the series of exhibitions in the Library of Congress honoring the centennials and sesquicentennials of American states and territories.

With the move of the Division from Capitol Hill to suburban Alexandria, adequate wall space was again available for creating a suitable exhibition area. In 1971, ten display boards were mounted along the east wall of the building's central corridor which links the reading room with the entrance foyer. Designed by Geography and Map Division staff members and built in the Library's carpenter shop, the boards permit the safe display of some 30 to 40 maps at a time.

In August 1971, the first exhibit was mounted: an exhibition of modern and historical maps of the Moon, celebrating the Apollo XV lunar mission. This was followed by substantive cartographic displays showing the development of the city of Alexandria, Virginia from colonial riverport to modern suburb of the Nation's Capitol, and an exhibit titled, "The World of Maps", which commemorated the 75th anniversary of the Geography and Map Division. The latter display included some 46 maps selected to illustrate the depth and variety of the Library's cartographic collections. The ten wall panels were subsequently augmented by a free standing display case placed in the Map Reading Room which permits the exhibition of three-dimensional cartographic objects such as atlases, globes and relief models.

Approximately one half dozen cartographic exhibits covering several periods and themes are displayed each year in the Pickett Street Annex. In 1977, these included maps of Africa, aviation cartography, Indians of North America, maps of Hawaii, and two showcase exhibits commemorating the 500th anniversary of the printing of Claudius Ptolemy's *Cosmographia* (1477) and the 400th anniversary of the death of the Italian publisher Antonio Lafreri.

The exhibition highlight of the past three decades was the display in the lower level gallery and corridors of the ground floor, Library of Congress building of "Maps for an Emerging Nation: Commercial Cartography in Nineteenth Century America". The exhibition was opened by Librarian of Congress Daniel J. Boorstin on August 7, 1977, to coincide with the convening in Washington, D.C. of the VIIth International Conference on the History of Cartography. The 112 item display, the result of the combined efforts of personnel from both the Library's Geography and Map Division and the Exhibits Office was supplemented by a 66-page exhibition catalog written by Walter W. Ristow.

Processing Activities

The story of map and atlas processing in LC's Geography and Map Division during the past three decades is essentially one dealing with the codification of cataloging philosophy and practice to high levels of sophistication: the development of MARC Map, the Library's machine-readable map cataloging system; the formulation and publication of resource tools and documentation ancillary to the growth of processing practice and application of machine systems; the maturation of the Division's annual Special Map Processing Project into an internationally recognized program of training and exchange for map librarians; and the sophistication of map and atlas preservation and storage techniques and equipment.

The MARC Map Project

From the initial step taken by the Library of Congress in 1941 appointing a full time map cataloger in the Geography and Map Division until 1968, atlas and map cataloging struggled along against the immense odds of the combination of too few staff and too many receipts.

A dramatic turnabout occurred when, in 1968, the Council on Library Resources was approached by Dr. Ristow and other division administrative staff with a proposal that G&M develop computer-assisted procedures for the bibliographic control of currently-received single and multi-sheet thematic maps, map series, and serially-issued maps. From the Council a grant was obtained and, with it, the seeds were sown for a revolution in map cataloging which is continuing yet today. Now dubbed MARC Map (MARC being the acronym for machine readable cataloging), the system was fully operable by late 1969. Modeled closely after the monograph machine-readable cataloging effort already in use by the Library's Processing Department, the system, identified those variable and fixed field data elements pertinent to map cataloging, developed sub-systems for recording and inputting the descriptive data, and, with assistance from personnel of the Library's Cataloging Distribution Service Division, formulated the parameters for multiple machine products generated through the system in the form of eye-readable catalog cards (VideoComp) and discrete or composite magnetic tapes. The latter services were introduced in April 1973 and have been on-going, without interruption or delay, ever since.

MARC Map, as of this writing, consists of a data file of over 55,000 formatted records of current and retrospective cartographic items. The file is increasing at a fairly uniform rate of about 7,000 titles per year.

Concomitant with the development of the machine-readable map cataloging system was the formulation of required system manuals and documentation. Two major tools developed thus far are the *MARC Map Data Preparation Manual* which outlines in detail, with descriptions and examples, the procedures to be followed in preparing map cataloging data in both variable and fixed-field positions for machine input. The *Manual* was first published in 1971; a second edition is in process. *Maps: A MARC Format*, a companion volume to the above, was first published in 1970. A second edition, published in 1976, incorporates a number of changes dictated largely by the growing sophistication of map cataloging both nationally and internationally, and the constantly changing state of the art of all computer-assisted bibliographic systems. The *Format*, providing a character-by-character analysis of map cataloging data as it appears on magnetic tapes, is used by systems analysts and software designers in developing alternative uses of the data beyond the traditional catalog cards, computer printout, either eye-readable or COM, and CRT-selected documentation.

MARC Map and/or its derivative systems is now in development or ongoing in five countries. Therefore, there is no doubt that it has irrevocably altered the nature of map processing systems. It has also provided the locus for all subsequent systems now operational or projected.

Map and Atlas Cataloging and Processing

During the period Walter Ristow was Chief of the Geography and Map Division, the concepts and essence of map and atlas cataloging underwent full metamorphosis. We will, however, only touch on the two most important areas.

One of the first published efforts systematically to arrange the world's geographic areas into a scheme useful for atlas classification occurred in 1928 with the appearance of the second edition of the Library of Congress' *Classification Class G*. A third edition in 1954 added maps and some additions and corrections were made in 1966. Never abandoned were the two special characteristics basic to the logic of the classification scheme, i.e., that geographic (areal) coverage has primacy over subject and that the final digit of any area form digit number formed a mnemonic constant.

Early in 1974 the Library's Subject Cataloging Division discussed with G&M staff members the desirability of a full revision of the schedule. Having determined that a comprehensive redrafting of the schedule was necessary, members of the Cataloging Unit began making changes to the schedule in five general areas: updating, which attempted to bring current geographic terminology into usage; correcting, which involved aspects of the printed schedule, making them agree with actual past and current cataloging practice, both at LC and throughout the North American map cataloging community; expanding the alphameric codes for second and third order administrative units, and subject subdivisions, which involved revising the hierarchical logic of the schedule resulting in vacating and/or relocating codes for certain geographic areas and redefining classification concepts and practices; and, lastly, the introduction of several outline maps to be used as aids to map and atlas classifications in general. This revision, accomplished as it was in a very short time, received recognition by the Library when, in March 1975, the Geography and Map Division staff members involved received meritorious service awards for their "outstanding and coordinated effort in preparing the revisions of the *Library of Congress Classification Class G* for maps and atlases".

Special Map Processing Project

The Geography and Map Division's annual Special Map Processing Project began in 1951 and has since grown and flourished as a most useful system for the mutual exchange of information, ideas, experience, and material between participants and Division personnel.

In the beginning the intent of the Project was to employ, temporarily, students and map librarians from participating institutions to search and subsequently process maps then held in the Division's large backlog of materials acquired during the post-War period. Most of the temporary employees were paid, but, even from the beginning, some members were co-operatively sponsored by their respective institutions. For the labor and services of their participants those institutions received upwards of several thousand maps, thus enriching their existing collections. There are no longer any participants supported by Library of Congress funds. Since 1972, the Project has been limited to persons supported by their institutions.

Over the years the Project has undergone other fundamental changes with regard to the nature of the sponsored participants and functions performed. Now, overwhelmingly, those candidates selected to represent participating academic institutions are graduate geographers and/or practicing map librarians, many with second masters degrees in Library Science, Information Science, or related disciplines. This is a marked departure from the early days when staffing of the project was accomplished almost entirely by hiring undergraduate students in geography.

The nature of the work has also changed, particularly during Walter Ristow's tenure as Division chief. Initially, all participants sorted and arranged backlogged material in the set map collection which benefitted solely the operations of the Division's Processing Section. Today, however, the members of the project are spread among the various administrative units and sections of the Division performing many varied and difficult tasks. Assignments for those tasks are made on the basis of completed questionnaires and other background information submitted by the participants and work priorities determined by Division supervisors.

More than 350 individuals representing some 110 institutions have contributed to the success of the project during the 28 years of its existence. It is estimated that close to 1.5 million maps and atlases have been selected by project participants for their sponsoring institutions. The Project has served and continues to serve as a significant acquisitions source for many of the participating institutions.

The continued success of the annual Special Map Processing Project has been due, in large part, to Walter Ristow, who has always maintained a keen interest in and concern for the growth of map and atlas collections in the United States and in map librarianship in general. Undoubtedly the confidence that map librarians have gained in the past decade, and the

Arch C. Gerlach, Chief Geographer, U.S. Geological Survey, presenting the National Atlas of the United States to L. Quincy Mumford, Librarian of Congress, Walter W. Ristow, Chief, Geography and Map Division, and Paul L. Berry, Director, Reference Dept., January 19, 1971.

general acceptance of maps and atlases as valued library reference tools are due to the success of the annual Special Projects under his guidance at LC's Geography and Map Division.

A c q u i s i t i o n s

As has been seen, there was no systematic map acquisitions program in the Library of Congress until the creation of a separate Division of Maps and Charts in 1897.

With the growth of the cartographic collections and staff increases over the years, the acquisitions program has undergone various changes (see Table I). A separate Acquisitions Section was organized in 1945. Following World War II, military agencies deposited in the Library of Congress selected issues of World War II maps. Map accessions for fiscal year 1945 accordingly passed 58,000, as compared with 7,700 in 1942. Following V-E Day (May 8, 1945) and V-J Day (September 2, 1945) there was rapid contraction of various wartime offices and agencies. Some map libraries, assembled to support various military and intelligence programs, were deactivated, and the Library's Map Division fell heir to their collections. Large military cartographic libraries weeded their holdings and transferred large quantities of maps and charts to the Library. More than 75,000 sheets were accessioned in fiscal 1947, and additional volumes of transferred material, which could not be processed by the small staff, were stored in the unfinished attic of the Annex building (now Thomas Jefferson Building).

Beside these transfers, the Division continued to keep abreast of the establishment, development and output of official, commercial, and private cartographic agencies throughout the world. The growth in map acquisition, which reflects accelerated map production during and following World War II, is particularly impressive.

Today acquisition activities are administered by the Division's Technical Services Section.

The Acquisitions Unit has responsibility for recommending cartographic and geographic items for acquisition by the Library. Coordinating its activities with the Exchange and Gift Division, the unit receives cartographic publications through government deposit from various federal, state, and local agencies, via the Copyright Office, and from individual donors. It cooperates with other federal map libraries and mapping agencies on the Interagency Map and Publications Acquisitions Committee to acquire foreign maps and atlases, by exchange or purchase. Personnel in the unit also have custody of the division's duplicate files, negotiate piece-for-piece and priced exchanges, and compile and maintain the division's acquisitions statistics and records.

The Geography and Map Division solicits information which may lead to acquisitions. In cooperation with the Library's Exchange and Gift Division, it welcomes bequests and gifts of maps and atlases which are not already in the collections. Recommendations are regularly made to the Library's Order Division for the purchase of items offered for sale by dealers or by private individuals.

Wartime demands had revealed significant gaps in the division's cartographic holdings, particularly in large- and medium-scale foreign topographic series. With the objectives of filling such gaps and reestablishing and expanding foreign exchanges, the Map Division joined with the Department of State and other federal cartographic agencies and libraries to establish a cooperative foreign map procurement program. The Interagency Map and Publication Acquisitions Committee (IMPAC) was organized the year that Dr. Ristow came to the Library. He soon implemented the Division's cartographic acquisitions program, coordinated through the Department of State which supported geographic attaches in several American embassies as well as a number of Washington-based cartographic specialists, who made periodic procurement missions to selected countries. For the past quarter century more than 90 percent of the Library's foreign cartographic accessions, both exchanges and purchases, have come via IMPAC channels.

S t a f f

As has been seen, in 1947 the Division had some 16 staff members. By 1957, the staff had increased to 18, by 1968 to 23, and at present it numbers 35. Virtually all of the growth has been due to the increase in technical processing activities which have taken place since 1968. Because of this, the Division was reorganized in 1973 into three general administrative units - the Office of the Chief; the Reference and Bibliography Section, and the Technical Services Section, which now includes the Acquisitions Unit, Cataloging Unit, Collections Maintenance Unit, and Data Preparation and Files Maintenance Unit. This section, under section head David K. Carrington, is the largest administrative unit within the Division.

Some eighty percent of current staff members hold bachelor's degrees, most of them in geography. Slightly less than half have graduate degrees. Included in the latter group are the forty percent of the staff who have masters degrees in library science - something rarely seen even a decade ago. Geography and map librarianship, to use the all-encompassing term for Division activities, is now a much more sophisticated albeit a bit more fragmented field, and all segments of the field are represented by the staff and what they do.

The staff/collection ratio has not improved much over the past eight decades, being now on the order of approximately 110,000 items per staff member. However, the new processing techniques have enabled us to gain better bibliographic control of the collections and that is certainly a great step forward.

P u b l i c a t i o n s

Since its inception, the Geography and Map Division has extended its reference services worldwide through an active publications program. As early as 1901, Philip Lee Phillips published his *List of Maps of America in the Library of Congress* which is still recognized as one of the classic reference volumes in cartobibliography. In 1909 Phillips issued the first two volumes of his *List of Geographical Atlases in the Library of Congress*, which is also a standard reference aid in historical cartographical research. Supplements were issued by Phillips in 1914 and 1920, and four additional volumes of *Geographical Atlases* were compiled by Clara Egli LeGear between 1958 and 1974.

The bibliography beginning on page 91 , compiled by Andrew M. Modelski, Division Bibliographer, is a complete list of officially published works of the Geography and Map Division beginning with Phillips' *Virginia Cartography ...*, issued by the Smithsonian Institution one year before the official establishment of the Division.

In recent decades Geography and Map Division specialists have compiled, and the Library has published, cartobibliographies on such varied subjects as aviation cartography, Civil War maps, treasure maps and charts, land ownership maps, three-dimensional maps, 19th century American railroad maps, ward maps of United States cities, and 19th century panoramic maps of Anglo-American cities. Currently in compilation, in cooperation with the American Revolution Bicentennial Office, is a comprehensive list of maps of the Revolutionary period in the Library of Congress.

Reference and Bibliography Section specialists have recently completed a checklist of the detailed fire insurance maps produced by the Sanborn Map Company, Pelham, New York, since 1867. The checklist, to be published by the Library of Congress, includes 10,000 entries recording 50,000 editions of insurance maps and atlases of American cities and towns, comprising an estimated 700,000 sheets.

The Geography and Map Division also provides bibliographical citations for the "Distinctive Recent Maps" feature of the American Congress on Surveying and Mapping's quarterly *Bulletin*; the *American Cartographer*; the Special Libraries Association, Geography and Map Division's *Bulletin*; the *Handbook of Latin American Studies*; and contributes citations to the international cartographic journals *Bibliographia Cartographica*; *Bibliographie Cartographique Internationale*; and *Imago Mundi*.

Walter W. Ristow, Chief,
Geography and Map Division, examining George Washington's
1749 manuscript map of Alexandria, Virginia.

S u m m a r y

We can call the period since 1950 one of great change - in equipment, materials, and technique. Cartography has changed into a much more sophisticated science and technology. Indeed, it has become a fully fledged discipline in which the practioner of map librarianship plays an increasingly important part.

The Division has led the way in establishing bibliographic control of cartographic materials through advances in classification, cataloging and also through the MARC Map program. During the past two decades the production of bibliographies and checklists by Division staff members has surpassed that of the early period. Reference work has also seen a gradual increase and is sure to make rapid advances one the Division relocates to its new home in the James Madison Memorial Building and is reunited with the staff and resources of the entire Library of Congress.

The Future

It is always difficult to predict with any great precision what will happen in the future. It has, however, been made quite evident by advances in mapping technology and information display, that great changes are imminent.

With the advent in the last few decades of entirely new systems of recording, synoptically, landscape changes throughout the world, e.g. remote sensing of the environment through aerial photography, thermal and radar imagery and satellite photography, a new and different look at processing controls must be taken. The rapid growth in computer mapping technics and in COM and other microimagery may very well require that map libraries acquire cartographic data in these varied forms and the equipment for its storage, retrieval and dissemination. All of this requires thorough research, investigation and experimentation, which will inevitably lead to a redefinition of the role of the map as an information storage device regardless of its format.

The future is certainly as exciting to us as it has been to our predecessors in the Division. We hope to be able to do as well as they have done in the care and nurturing of the products of this ancient and venerable graphic art and science. We are grateful for the chance to continue and build upon the marvelous legacy our pioneering and energetic forebearers have bequeathed to us.

References

JOHNSTON, William Dawson: In: ... *History of the Library of Congress*. Vol. 1, 1800-1864. Washington 1904.

LEGEAR, Clara Egli: Early Years in the Map Division, Library of Congress. Special Libraries Association, Washington, D.C. Chapter, Geography and Map Group. In: *Federal Government Map Collecting; A Brief History*. Washington 1969. p.20-28.

MEARNS, David C.: In: *The Story Up to Now: The Library of Congress 1800-1946*. Washington: Library of Congress 1949. 226 p.

MODELSKI, Andrew M.: 100 Years of Cartographic Bibliography; The Library of Congress' Bibliography of Cartography. Paper presented at the 69th Annual Conference, Special Libraries Association, Kansas City, June 1978. [Typescript] 15 p.

RISTOW, Walter W.: Cartographic Exhibits. In: *Surveying and Mapping*. 14, 1954, 1. p.18.

RISTOW, Walter W.: Cartographic Information Services of the Library of Congress. In: *The American Cartographer*. 1, 1974, 2. p.125-130.

RISTOW, Walter W.: Philip Lee Phillips, Cartobibliographer. In: *Karten in Bibliotheken*. Festgabe für Heinrich Kramm zur Vollendung seines 65. Lebensjahres. - Bonn-Bad Godesberg 1971. p.95-109. (Kartensammlung und Kartendokumentation. 9.)

RISTOW, Walter W.; David K. CARRINGTON: Machine Readable Map Cataloging in the Library of Congress. In: *Special Libraries* 62, 1971, 9. p.343-352.

STEPHENSON, Richard W.: Congress' First Map Collection. Special Libraries Association, Washington, D.C. Chapter, Geography and Map Group. In: *Federal Government Map Collecting; A Brief History*. Washington 1969. p.7-19.

STEPHENSON, Richard W.: Reference Facilities and Services of the Geography and Map Division, Library of Congress. In: *Special Libraries*. 65, 1974, 5/6. p.227-233.

U.S. Library of Congress. In: *Annual Report of the Librarian of Congress*. Washington 1897-

U.S. Library of Congress, Geography and Map Division. Unpublished annual reports of the Division. 1897-

WOLTER, John A.: Geographical Libraries and Map Collections. In: *Encyclopedia of Library and Information Science*. Vol. 9. - New York: Marcel Dekker 1973. p.236-266.

TABLE I

Geography and Map Division Accessions and Holdings

Fiscal Year	Accessions[1]	Holdings			
		Maps	Atlases	Other[2]	Total
1898	736	48,605	1,180	410	50,195
1917	9,709	439,405	5,059	2,439	446,903
1937	15,717[4]	1,376,801	11,684	4,509	1,392,994
1957[3]	28,251[4]	2,322,141	18,000	1,000	2,341,141
1977	19,308[4]	3,593,706	40,097	1,235	3,635,038

[1] *During fiscal year includes all materials added to the collections.*

[2] *Included at different times - books, globes, manuscripts, views, etc. From 1957 on only globes, models, and other cartographic formats.*

[3] *Estimate except for total.*

[4] *Net increase.*

TABLE II

Geography and Map Division Staff, Space Occupied and Location

Fiscal Year	Staff	Space Occupied (sq.ft.)	Location
1897	3	10,000	(LCB) LC Main Building
1917	6[1]	10,000	(LCB) " " "
1937	8[1]	15,000[2]	(LCB) " " "
1957	18	47,000	(TJB) LC Annex Building
1977	35	75,000	(PSA) Pickett St. Annex Alexandria, Va.
1980[3]	40[2]	93,000	(JMMB) James Madison Memorial Building

[1] *Average number varied during reporting period.*

[2] *Estimate.*

[3] *Planned layout - move from PSA to JMMB during fiscal year 1980.*

TABLE III

Use of G&M Division Collections

Fiscal Year	Readers Assisted	Telephone	Correspondence	Total
1902	789[1]	-	122	911
1917	2,336[2]	-	644	2,980
1937	5,000[3]	-	1,041	6,041
1957	8,263	2,103	973	11,339
1977	8,251	5,234	3,857	17,342

[1] *First year for which adequate statistics are available.*

[2] *Estimate based on three month count.*

[3] *Estimate based on circulation.*

LIST OF OFFICIALLY PUBLISHED WORKS OF THE GEOGRAPHY AND MAP DIVISION

Compiled by

Andrew M. MODELSKI

1896

... Virginia cartography; a bibliographical description, by P. Lee PHILLIPS. City of Washington: Smithsonian institution, 1896. 85 p. (Smithsonian miscellaneous collection. Vol. XXXVII, art. IV). Publication 1039.

1898

Guiana and Venezuela cartography. By P. Lee PHILLIPS. Excerpt from: American Historical Association. Annual Report for the year 1897. Washington: Govt. Print. Off.,1898. p.681-776. 96 p.

Alaska and the northwest part of North America, 1588-1898; maps in the Library of Congress, by P. Lee PHILLIPS. Washington: Govt. Print. Off., 1898. 119 p.

List of books relating to Cuba (including references to collected works and periodicals), by A.P.C. GRIFFIN ... with a bibliography of maps by P. Lee PHILLIPS... Washington: Govt. Print. Off., 1898. 61 p. (U.S. 55th Cong., 2nd sess., 1897-98. Senate. Doc. no. 161).

1900

List of maps and views of Washington and District of Columbia in the Library of Congress, by P. Lee PHILLIPS. Washington: Govt. Print. Off., 1900. 77 p. (56th Cong., 1St sess. Senate Doc. 154.)

1901

A list of maps of America in the Library of Congress, preceded by a list of works relating to cartography. By P. Lee PHILLIPS. Washington: Govt. Print. Off., 1901. 1137 p.

A list of works relating to cartography. By P. Lee PHILLIPS. Washington; Govt. Print. Off., 1901. 90 p. (Reprinted from "A list of maps of America, in the Library of Congress".)

A list of books, magazine articles, and maps relating to Brazil. 1800-1900. Prepared by P. Lee PHILLIPS ... A supplement to the handbook of Brazil (1901) compiled by the Bureau of the American republics. Washington: Govt. Print. Off., 1901. 145 p.

1902

A list of books, magazine articles, and maps relating to Central America, including the republics of Costa Rica, Guatemala, Honduras, Nicaragua, and Salvador. 1800-1900. Prepared by P. Lee PHILLIPS ... Comp. for the Bureau of the American republics. Washington: Govt. Print. Off., 1902. 109 p.

1903

A list of books, magazine articles, and maps relating to Chile. Prepared by P. Lee PHILLIPS ... Comp. for the International Bureau of the American republics. Washington: Govt. Print. Off., 1903. 110 p.

A list of books (with references to periodicals) on the Philippine islands in the Library of Congress. By A.P.C. GRIFFIN ... with chronological list of maps in the Library of Congress, by P. Lee PHILLIPS ... Washington: Covt. Print. Off., 1903. 397 p.

1904

Check list of large scale maps published by foreign governments (Great Britain excepted) in the Library of Congress. Comp. under the direction of Philip Lee PHILLIPS. Washington; Govt. Print. Off., 1904. 58 p.

The Kohl collection (now in the Library of Congress) of maps relating to America. By Justin WINSOR. A reprint of Bibliographical contribution number 19 of the Library of Harvard University. With index by Philip Lee PHILLIPS. Washington: Govt. Print. Off., 1904. 189 p.

1905

Maps relating to Cuba [1492-1905]. By P.L. PHILLIPS. Reprint from: International Bureau of the American Republics. Washington: Govt. Print. Off.,1905. 66 p.

1909

A list of geographical atlases in the Library of Congress, with bibliographical notes. Comp. under the dir. of Philip Lee PHILLIPS. Washington: Govt. Print. Off.,1909. 2 vol., 1659 p.

1912

The Lowery collection. A descriptive list of maps of the Spanish possessions within the present limits of the United States, 1502-1820; by Woodbury Lowery. Ed. with notes by Philip Lee PHILLIPS. Washington: Govt. Print. Off., 1912. 567 p.

1914

Author list of the geographical atlases in the Library of Congress ... Comp. under the dir. of Philip Lee PHILLIPS. Washington: Govt. Print. Off., 1914. 137 p. (Reprinted from the third volume of the List of geographical atlases.)

A list of geographical atlases in the Library of Congress, with bibliographical notes. Comp. under the dir. of Philip Lee PHILLIPS. Washington: Govt. Print. Off.,1914. Vol. 3. 1030 p.

1915

Notes on the cataloging, care, and classification of maps and atlases, including a list of publications compiled in the Division of maps and charts. By Philip Lee PHILLIPS. Washington; Govt. Print. Off., 1915. 20 p. ("These 'Notes', originally contributed to the fourth edition of Charles A. Cutter's 'Rules for a dictionary catalog', have since been amplified.")

1918

A list of atlases and maps applicable to the world war; comp. under the dir. of Philip Lee PHILLIPS. Washington: Govt. Print. Off., 1918. 202 p.

1920

Author list of the geographical atlases in the Library of Congress ... Enl. and rev. ed. Comp. under the dir. of Philip Lee PHILLIPS. Washington; Govt. Print. Off., 1920. 163 p.

A list of geographical atlases in the Library of Congress, with bibliographical notes. Comp. under the dir. of Philip Lee PHILLIPS. Washington: Govt. Print. Off.,1920. Vol. 4. 639 p.

1921

Notes on the cataloging, care, and classification of maps and atlases, including a list of publications compiled in the Division of Maps. Rev. ed. by Philip Lee PHILLIPS. Washington: Govt. Print. Off., 1921. 21 p. ("These 'Notes', originally contributed to the fourth edition of Charles A. Cutter's 'Rules for a dictionary catalog', have since been amplified".)

1924

Notes on the life and works of Bernard Romans. By Philip Lee PHILLIPS. Deland, Florida; Florida State Historical Society 1924. 128 p. (With a reproduction in two parts, of Romans' "Part of the Province of East Florida".)

1926

A descriptive list of maps and views of Philadelphia in the Library of Congress 1683-1865. By P. Lee PHILLIPS. [Philadelphia]: Geographical Society of Philadelphia 1926. 91 p. ("Issued by permission of the Library of Congress, as Special Publication number two of the Geographical Society of Philadelphia, 1926".)

1928

Classification. Class G: Geography, anthropology, folklore, manners and customs, sports and games. 2nd ed. Washington 1928. 180 p.

Noteworthy maps; accessions, with acknowledgment of sources of gifts, exchanges, and transers. no.[1]-3; 1925/26-1927/28. Washington: U.S. Govt. Print. Off. 95 p. ("A continuation of the lists included in the Annual reports of the Librarian of Congress for the years 1899-1925".)

1935

Catalogue of an exhibition in the Division of maps comprising some 200 Hispanic-American maps, atlases, geographies, globes, and portraits of historical, diplomatic, and cartographic interest, ranging through four centuries. By Lawrence MARTIN, Edith FITTON and Clarence G. JOHNSON. Upon the occasion of the general assembly of the Pan American institute of geography and history at Washington D.C., October 14 to 19, 1935. [Washington, D.C.:] U.S. Govt. Print. Off., Library branch, 1935. 20 p.

1937

Disturnell's map. By Lawrence MARTIN. Washington: U.S. Govt. Print. Off., 1937. 33 p., ill. [facsim.] ("Study of an important American treaty map ... made in the Library of Congress at the request of the Department of state.")

1940

Catalogue of an exhibition of maps in National academy of sciences building ... comprising some 400 maps and other materials related to the programs of section III, geological sciences, and section VIII, history and geography. Arranged by Lawrence MARTIN. Upon the occasion of the Eighth American scientific congress at Washington, D.C., May 10-18, 1940. Washington: U.S. Govt. Print. Off. (Library branch) 1940. 14 p.

1946

Classification. Class G: Maps G 3160-9999. [Preliminary Draft] Washington 1946. 58 p.

1949

Maps; their care, repair and preservation in libraries. By Clara Egli LEGEAR. Washington 1949. 46 p.

A supplementary key to accompany the V.K. Ting atlas of China (edition of 1934). Comp. by L.W. WADSWORTH. Washington 1949. 22 p., maps.

1950

Maps; their care, repair and preservation in libraries. By Clara Egli LEGEAR. Washington 1950. 46 p.

Library of Congress Departmental & Divisional Manuals. No. 15. Map Division. Washington 1950. 37 p.

United States atlases; a list of national, State, county, city and regional atlases in the Library of Congress. Comp. by Clara Egli LEGEAR. Washington 1950. 445 p.

1951

The Hotchkiss map collection; a list of manuscript maps, many of the Civil War period, prepared by Major Jed. Hotchkiss, and other manuscript and annotated maps in his possession. Comp. by Clara Egli LEGEAR, with a foreword by Willard Webb. Washington 1951. 67 p.

Marketing maps of the United States; an annotated list. Comp. by Walter W. RISTOW. Washington 1951. 54 p.

The services and collections of the Map Division. By Walter W. RISTOW. Washington: Library of Congress 1951. 22 p., ill., maps.

Three-dimensional maps; an annotated list of references relating to the construction and use of terrain models. Comp. by Walter W. RISTOW. Washington 1951. 40 p.

1952

International Geographical Congress. *17th, Washington, D.C., 1952*. Catalog of national exhibits. Ed. by Charles W. BUFFUM. Washington 1952. 81 p.

Marketing maps of the United States; an annotated bibliography. 2nd rev. red. Comp. by Marie C. GOODMAN and Walter W. RISTOW. Washington 1952. 102 p.

1953

United States atlases; a catalog of national, state, county, city, and regional atlases in the Library of Congress, and cooperating libraries. Comp. by Clara Egli LEGEAR. Washington 1953. 301 p. (Issued as suppl. to the Library of Congress catalog.)

1954

Classification. Class G: Geography, anthropology, folklore, manners and customs, recreation. 3rd ed. Washington 1954. 502 p.

Classification. Class G: Geography, anthropology, folklore, manners, and customs, recreation. 3rd ed. (with supplementary pages) Washington: U.S. Govt. Print. Off., 1954; (repr. 1966.) 579 p.

A guide to historical cartography; a selected, annotated list of references on the history of maps and map making. Comp. by Walter W. RISTOW and Clara E. LEGEAR. Washington 1954. 14 p.

1955

Library of Congress catalog; a cumulative list of works represented by Library of Congress printed cards. Maps and atlases. Washington 1953-(55). (3) vol. 137 p.

Treasure maps in the Library of Congress; an annotated list. Comp. by J. Douglas HILL and Richard S. LADD. Washington 1955. 24 p.

1956

Aviation cartography; a historico-bibliographic study of aeronautical charts. By Walter W. RISTOW. Washington 1956. 114 p.

Aviation cartography; a historico-bibliographic study of aeronautical charts. By Walter W. RISTOW. Washington 1956. 114 p. (Repr. by U.S. Aeronautical Chart and Information Center.)

Maps; their care, repair and preservation in libraries. By Clara Egli LEGEAR. Rev. ed. Washington 1956. 75 p.

1957

Captain John Smith's map of Virginia, Facsimile of an engraving in the collections of the Library of Congress. Accompanying text by Walter W. RISTOW. Washington 1957. 4 p.

Library of Congress Departmental & Divisional Manuals. No. 15. Section B. Map Processing. Map Division. Washington 1957. 36 p.

Treasure maps in the Library of Congress; an annotated list. Comp. by J. Douglas HILL and Richard S. LADD. Rev. ed. Washington 1957. 24 p.

1958

A list of geographical atlases in the Library of Congress, with bibliographical notes. (A continuation of four volumes by Philip Lee Phillips). Comp. by Clara Egli LEGEAR. Library of Congress, Geography and Map Division. Washington 1958. Vol. 5. 666 p.

Marketing maps of the United States; an annotated bibliography. 3rd (rev.) ed. Comp. by Walter W. RISTOW. Washington 1958. 147 p.

1959

Selected maps and charts of Antartica; an annotated list of maps of the South Polar regions published since 1945. Comp. by Richard W. STEPHENSON. Washington 1959. 193 p.

1960

Aviation cartography; a historico-bibliographic study of aeronautical charts. By Walter W. RISTOW. 2nd ed. Rev. and enl. Washington 1960. 245 p.

Facsimiles of rare historical maps available for sale. Comp. by Walter W. RISTOW. Washington 1960. 6 p.

A guide to historical cartography; a selected, annotated list of references on the history of maps and map making. Comp. by Walter W. RISTOW and Clara E. LEGEAR. 2nd ed., rev. Washington 1960. 22 p.

1961

Civil War maps; an annotated list of maps and atlases in map collections of the Library of Congress. Comp. by Richard W. STEPHENSON. Washington 1961. 138 p.

1962

Maps showing explorers' routes, trails & early roads in the United States; an annotated list. Comp. by Richard S. LADD. Washington: U.S. Govt. Print. Off. 1962. 137 p.

1963

A list of geographical atlases in the Library of Congress, with bibliographical notes. (A continuation of four volumes by Philip Lee Phillips.) Comp. by Clara Egli LEGEAR. Library of Congress, Geography and Map Division. Washington 1963. Vol. 6. 681 p.

1964

A descriptive list of treasure maps and charts in the Library of Congress. Comp. by Richard S. LADD. Washington 1964. 29 p.

1965

Three-dimensional maps; an annotated list of references relating to the construction and use of terrain models. Comp. by Walter W. RISTOW. 2nd ed., rev. and enl. Washington: U.S. Govt. Print. Off. 1964 [i.e. 1965] 38 p.

1966

Facsimiles of rare historical maps, a list of reproductions for sale by various publishers and distributors. Comp. by Walter W. RISTOW. Rev. Washington 1966. 12 p.

1967

Land ownership maps, a checklist of nineteenth century United States county maps in the Library of Congress. Comp. by Richard W. STEPHENSON. Washington: U.S. Govt. Print. Off. 1967. 86 p., ill., maps.

1968

Detroit and vicinity before 1900; an annotated list of maps. Comp. by Alberta G. AURINGER KOERNER. Washington: Library of Congress 1968. 84 p., maps.

Facsimiles of rare historical maps, a list of reproductions for sale by various publishers and distributors. Comp. by Walter W. RISTOW, assisted by Mary E. GRAZIANI. 3rd. ed., rev. and enl. Washington 1968. 20 p.

1970

Maps, a MARC format; specifications for magnetic tapes containing catalog records for maps. [Final document prepared by Patricia E. Parker]. Washington: U.S. Govt. Print. Off. 1970. 45 p.

1971

Alexandria: City and Suburb. An exhibition of maps dating from 1745 to 1970. Prepared by Richard W. STEPHENSON and Joan ENERSON. Washington 1971. 15 p.

Data preparation manual for the conversion of map cataloging records to machine-readable form. Prepared by David K. CARRINGTON and Elizabeth U. MANGAN. Washington: Library of Congress 1971. 317 p.

1972

Captain John Smith's map of Virginia, Facsimile of an engraving in the collections of the Library of Congress. Accompanying text by Walter W. RISTOW. Text rev. and repr. Washington 1972. 4 p.

The World of Maps. An exhibition of maps commemorating the 75th anniversary year of the Geography and Map Division. Prep. by Richard W. STEPHENSON, Andrew M. MODELSKI and John R. HÉBERT. 1972. 32 p.

1973

The bibliography of cartography. Boston; Hall 1973. 5 vol. 3393 p. [A reproduction of a card bibliography in the Geography and Map Division, which contains an estimated 90,000 entries for works published from the early 19th century through 1971.]

A descriptive list of treasure maps and charts in the library of Congress. 2nd ed. Comp. by Donald A. WISE. Washington: U.S. Govt. Print. Off. 1973. 30 p., ill.

Guide to the history of cartography; an annotated list of references on the history of maps and mapmaking. Comp. by Walter W. RISTOW. Washington: Geography and Map Division, Library of Congress 1973. 96 p.

A list of geographical atlases in the Library of Congress, with bibliographical notes. (A continuation of four volumes by Philip Lee Phillips.) Comp. by Clara Egli LEGEAR, Library of Congress, Geography and Map Division. Washington 1973. Vol. 7. 708 p.

Maps of the Mississippi River. An exhibition of maps commemorating the discovery and exploration of the Mississippi River by Marquette and Joliet in 1673. Prep. by Richard W. STEPHENSON, Andrew M. MODELSKI and John R. HÉBERT. 1973. 28 p.

1974

A list of geographical atlases in the Library of Congress, with bibliographical notes. (A continuation of four volumes by Philip Lee Phillips). Comp. by Clara Egli LEGEAR. Washington: Library of Congress, Geography and Map Division 1974. Vol. 8. 109 p.

Panoramic maps of Anglo-American cities; a checklist of maps in the collections of the Library of Congress, Geography and Map Division. Comp. by John R. HÉBERT. Washington: Library of Congress 1974. 118 p., ill.

1975

The Geography and Map Division: a guide to its collections and services. Rev. ed. [by Walter W. RISTOW]. Washington; Library of Congress 1975. 42 p.

Railroad maps of the United States; a selective annotated bibliography of original 19th-century maps in the Geography and Map Division of the Library of Congress. Comp. by Andrew M. MODELSKI. Washington: Library of Congress 1975. 112 p.

Ward maps of United States cities: a selective checklist of pre-1900 maps in the Library of Congress. Comp. by Michael H. SHELLEY. Washington: Library of Congress 1975. 24 p.

1976

Classification, class G: geography, maps, anthropology, recreation / Subject Cataloging Division, Processing Department, Library of Congress. 4th ed. Washington: Library of Congress 1976. 435 p., maps.

Maps, a MARC format: specifications for magnetic tapes containing catalog records for maps / MARC Development Office. 2nd ed. [prepared by Patricia PARKER and Elizabeth MANGAN]. Washington: Library of Congress 1976. 76 p.

1977

Maps for an emerging nation, commercial cartography in Nineteenth-century America; an exhibition at the Library of Congress. By Walter W. RISTOW. Washington: Library of Congress 1977. 66 p.

Nautical charts on vellum in the Library of Congress. Comp. by Walter W. RISTOW and R.A. SKELTON. Washington: Library of Congress 1977. 31 p.

MAP LIBRARIANSHIP COMES OF AGE

By Helen Wallis

The American historian Justin Winsor, librarian of Harvard University Library, in 1884 appointed a map curator with the encouraging words: "Well, all I can do is to turn you loose and let you flounder". The curator in question, H.C. Badger, published an account of his experiences in the *Library Journal*, September 1892, under the title "Floundering among the Maps". The phrase well describes the activities of curators and readers in an unorganized map collection. Up to 1950 most map librarians had to be selfeducated, and only too often they found that the needs of map collections were low in the list of priorities as conceived by library administrators. Thus Walter Ristow points out in his paper "The Greening of Map Librarianship", which was published by the Special Libraries Association, Geography and Map Division Bulletin, no. 111, 1978, that when he started work as Map Librarian of New York Public Library in 1937 there were fewer than 30 map libraries in the United States with one or more full-time employees. Of these, 25 per cent were in university libraries, a third were in public libraries and some 20 per cent in Federal agencies.

In Europe many map collections remained unexploited up to the second World War for lack of trained staff and proper catalogues. The British Museum had published its *Catalogue of Printed Maps, Plans and Charts* in two volumes in 1885. When a new edition of the catalogue was published in 1967, it could still claim to be the only national library to have published a complete map catalogue. Co-ordinated work in recording collections and establishing standards was still in a rudimentary form at the opening of the post-war period. What gave impetus to the movement for establishing map librarianship on a professional footing was the great increase in modern map production during the War and the large consignments of cartographic material passed to libraries for preservation after the second World War. The more extensive use of map collections by public officials had also put the map library, so to speak, on the map. The United States can claim precedence in the New World, and perhaps overall, in its establishment of a map librarians' group within a national library association. The Geography and Map Group of the Special Libraries Association was established in the Washington chapter in October 1941, and in 1944 it was promoted to the status of a Division with a membership of about 50 people. Walter Ristow himself with Clara E. Le Gear as Division Chairman went on to organize in 1945 a group in New York City, and this held its first meeting on 3 May 1945. The *Bulletin* of the Group started as a four-page newsletter and developed into a quarterly of some 80 pages. Many important papers on map librarianship have appeared between its covers, and it provides a fine documentation of the development of map librarianship in the USA during the last 35 years. A special One Hundredth Issue which appeared in June 1975 carried an impressive account of the Division's progress. The Division also published *Map Collections in the United States and Canada* (1954, second edition 1970, third edition 1977).

In Great Britain no special group has been set up in the Library Association. At the annual Symposium of the British Cartographical Society held at Swansea in 1965, a half-day session entitled "What is a Map Library" and provoked from R.A. Skelton, then Superintendent of the Map Room of the British Museum, the proposal (which he described as a hobby-horse of his) that map curators should form a group or circle within an existing organization. This was followed up by a meeting of map curators convened at the British Museum in June 1966. It was decided to form such a group, and the appropriate body chosen as parent was the British Cartographic Society, founded three years earlier. The Map Curators Group of the Society held its first formal meeting at the Royal Geographical Society on 8 December 1966. It was agreed to hold about twice a year regular meetings in the form of visits to map collections, in addition to half-day sessions fitted into the annual Symposia. News of the activities of the Group is given in the British Cartographic Society's publication, the *Cartographic Journal,* issued twice a year. The Group has now visited many of the major map collections of the country. It extended its activities to education and training when jointly with ASLIB it held in London from 1969 a series of annual courses on "Work with Maps", some aimed at beginners, others for established map librarians on particular themes, such as "Work with Older Maps" (1974) and "The Management of Map Collections" (1975). From 1976 the National Library of Scotland jointly with Edinburgh University Library and the British Cartographic Society has run similar courses. Both these series of courses have been very well attended and clearly meet a serious need in the profession.

An organization complementary to the British Cartographic Society, the Society of University Cartographers, was founded in 1964 to cater for cartographers and map curators in departments of geography in the universities and technical colleges. This also holds annual seminars and its *Bulletin* is issued twice a year.

Differing circumstances from one country to another explain why a variety of answers has been provided to the question: to which parent organizations, if any, map librarians should attach themselves. In Canada, the Association of Canadian Map Librarians was established in 1967, one year after the Western Association of Map Libraries had provided North America with a regional group for the Pacific seaboard. Both these are independent. They each run excellent publications: the *ACML Bulletin*, from 1968 and the *WAML Information Bulletin*, from 1969, and they published in 1969 Directories of Map Collections, the Canadian one being followed by a second edition in 1975.

Elsewhere in the world the same diversity is found. The Netherlands and Denmark both have Cartographical Societies within which map librarians find a forum for discussion. The Australian Map Curator's Circle was founded in 1973 and holds annual seminars, whose proceedings are printed in *The Globe*. Its publications include *A Directory of Map Collections in Australia* (1974). A New Zealand Mapkeepers Circle has been founded and held its first seminar in February, 1977, its second at Hamilton in February 1978. News of these proceedings was published in *The Globe*, no. 8 (1977).

South Africa at present has no comparable organization but is planning to hold in the summer of 1979 a seminar or workshop for map curators.

The establishment of the Geography and Map Libraries Subsection in the International Federation of Library Associations in 1969 was a major landmark in the development of map librarianship. This event took place at the conference of IFLA at Copenhagen in August 1969 after a year of careful planning under the guidance of Karl A. Baer, Chairman of the Special Libraries Division of IFLA. Walter Ristow was elected Chairman and Helen Wallis, Secretary. The first official meeting of the Subsection was held at the IFLA Conference at Moscow in 1970. In accordance with a proposal that the Subsection should meet at first only in even-numbered years, the next meetings were at Budapest in 1972, and Washington D.C. in 1974. The Subsection participation in the discussions on the restructuring of IFLA took members of its Standing Committee and others to meetings at Oslo in 1975 and Lausanne in 1976. The 50 year Jubilee meeting at Brussels in 1977 was once again an open meeting with a large attendance, and the Subsection now achieved full recognition by election to the status of a Section in the Special Libraries Division. The proceedings of the Subsection (or Section) are published in *Inspel*, the official organ of the Special Libraries Division.

The project taken up by the Subsection in 1969 as its first task was the compilation of a *World Directory of Map Collections*, under the direction of Walter Ristow, with contributions from libraries all over the world. The volume was published by Verlag Dokumentation, München, in time for the IFLA meeting in Lausanne in August 1976. A second edition will be prepared for publication in about 1984. The Section has now set up further Working Groups. One for the training of map librarians, Chairman Lothar Zögner, has the following terms of reference:

(a) to investigate the current situation in respect of the training of map librarians and to make recommendations;

(b) to prepare a Practical Manual for Map Librarianship;

(c) to hold seminars.

Another is preparing an International Glossary of Cartographic Terms, under the joint Chairmanship of A.V. Kozlova (USSR) and D.K. Carrington (USA). The Working Party as a secondary task maintains liaison with the ICA Commission on the Multilingual Dictionary of Technical Terms in Cartography. More recently a UNIMARC Cartographic Materials Working Group has begun work, with Hugo Stibbe as Chairman. The Working Group for the preparation of the second edition of the World Directory of Map Collections is under the Chairmanship of John Wolter (USA) and will start work in due course. Meanwhile, corrections and additions to the First Edition are welcome as preparatory material for inclusion in the second.

Late in 1973 the Geography and Map Libraries Subsection proposed the establishment of a Joint Working Group with the IFLA Committee on Cataloguing to draft an International Standard Bibliographic Description for Maps. The work was organized by IFLA's International office for Universal Bibliographic Control, and Hugo L.P. Stibbe entrusted with the Chairmanship of the Working Group. Their recommendations, *ISBD(CM): International Standard Bibliographic Description for Cartographic Materials* were published by the IFLA International office for U.B.C. in 1977.

Notable changes have also taken place in map cataloguing in recent years. The MARC Map Unit of the Library of Congress established in 1968 has been the pioneer in the development of a machine-readable format for maps. As Thomas Parr points out in his valuable paper "Automation of Cartobibliography: review of MARC for map library information retrieval and cartographic bibliography", *S.L.A. Geography and Map Division Bulletin*, No. 100 (1975), p.43, MARC now represents an evolving internationally accepted standard, supported by major government programmes in the U.S.A., Canada, Great Britain, France, Germany and the Netherlands. For maps the major problem is multiple-level cataloguing, and this has not yet been solved, although the Library of Congress is now incorporating into its system a process of two level cataloguing.

Hans van de Waal describes in his essay in this volume the application of a two level MARC process to the preparation of the Dutch *Bibliografie van in Nederland Verschenen Kaarten, 1975* (1977). From 1979 each previous year's bibliography will be published annually. The Bureau of the Dutch Union Map Catalogue (DUMC), which is housed in the Geografisch Instituut in Utrecht, receives the maps by a system of voluntary deposit. In Canada a National Union Catalogue Committee has been set up with the ultimate aim of producing a national bibliography of map cataloguing records, since *Canadiana,* the national bibliography, does not include maps. Work on the Canadian cataloguing rules for maps is in the charge of Hugo Stibbe, whose report *MARC-maps: the history of its development and a current assessment* (1976) is an important addition to the literature.

In Great Britain, as in Canada, the national bibliography does not yet include sheet maps, although it does handle atlases. A Working Party has now been set up by the British Library to investigate the feasibility of producing a record by automation to include all sheet maps, and it is hoped that this can be linked with a proposal of the Ordnance Survey for an automated cumulative record of their map publications. The development of BLAISE, the British Library automated information system, would make the British National Bibliography for maps readily available throughout the country. The introduction of an automated system would also involve the Map Library's own cataloguing system, at present on traditional lines and produced in volume form. The one volume *British Library Catalogue of Printed Maps, Charts and Plans: Ten-year Supplement, 1965-74* was published in the spring of 1978. This brings almost up to date what was formerly described as the British Museum Catalogue. It should be explained that in 1973 all the library departments of the British Museum were transferred to a new organization, the British Library; hence the change of name. Innovations can also be reported from Scotland. The National Library of Scotland has provided the current Bibliography of Scotland with data of all relevant map publications since January 1976. Although these are now in the computer, they will not appear in volume I because of their bulk, but they await eventual publication. A greater uniformity in the recording of maps has been achieved through the work of a panel of the British Standards Institution whose membership was drawn from the major map libraries. Its report, *Recommendations for bibliographical references to maps and charts, B.S.5195. Part I. References in accessions lists,* was published in 1975, and *Part II. References in Books and Articles,* in 1977. Libraries such as the Bodleian in its duplicated accession lists and the Royal Geographical Society in its *New Geographical Literature and Maps* published annually, use the recommended system to good effect. Correspondence in the January, February and March issues of the *Geographical Magazine* (1978), describe these various developments in Canada, Great Britain and the Netherlands. For Great Britain a section on Map Libraries (by Helen Wallis) was included in *British Librarianship and Information Science 1971-1975* (1977); another indication that map librarianship is now an acknowledged part of the library system.

Carto-bibliographies are invaluable tools for the map library, and their compilation may be regarded as an appropriate part of a map librarian's duties. A regional project in co-operative cataloguing has been initiated by the Hermon Dunlap Smith Center for the History of Cartography established in 1972 at the Newberry Library in Chicago. This is the Midwest Map Cataloguing Project for the production of a union catalogue of pre-1900 maps of America in midwestern libraries and historical societies. An international project of earlier vintage is the compilation of a union catalogue of maps, AD1200-1500, set up by the International Geographical Union at its 13th Congress, Paris, 1931, under the Chairmanship of the late Professor Roberto Almagiá. The first volume, *Mappemondes,* edited by Marcel Destombes, was published in 1964 to coincide with the 20th International Congress in London. The project is now in the hands of a Working Group under the Chairmanship of George Kish, and the volume for printed maps is in preparation.

Notable examples of cartobibliographies achieved through individual initiative, are Cornelis Koeman's *Atlantes Neerlandici,* 5 volumes, (1967-71), R.A. Skelton's *County atlases of the British Isles, 1579-1703,* (Carta Press, 1970), and the important series published by the Library of Congress. The publication of the card catalogues of major libraries in volume form has likewise added to the corpus of information generally available. Current bibliographies include the *Bibliographie Cartographique Internationale* begun in 1950 as a listing of map publications, and the *Bibliographia Cartographica*, from 1974, formerly *Bibliotheca Cartographica* (1957-1972), comprising lists of the current cartographical literature. *Imago Mundi: A Review of Early Cartography,* remains the only learned international journal on the history of cartography. It was founded as an annual publication in 1935 by Leo Bagrow and in 1975 was taken over by a private non-profit making company registered in London. The volumes issued annually are now designated the Journal of a newly formed Society operating under the Board of Imago Mundi and called the International Society for the History of Cartography. Thus Volume 27 of *Imago Mundi* is also enumerated Second series, Volume 1. Another useful handbook for map curators is the Directory of Research Workers in the History of Cartography, prepared by Professor E.M.J. Campbell, Professor of Geography of Birkbeck College, London in collaboration with Mr. P.K. Clark.

The International Geographical Congresses have shown interest in cartography since the time of the first Congress in 1871 at Antwerp, which featured a fine exhibition of early maps. In 1958 in response to the need for a more extensive exchange of ideas among cartographers, the International Cartographic Association was founded in association with the International Geographical Union. Its meetings are held every two years, and its regular publication is the *International Yearbook of Cartography.* A Working Group on the history of cartography was established by the I.C.A. in 1972, and in 1976 was promoted to the status of a Commission, with the project of compiling an historical glossary of cartographic innovations and their diffusion. Many map librarians are numbered among contributors to this glossary, which is due for publication in time for the International Conference of the ICA at Tokyo in 1980.

Another offshoot of the International Geographical Congress is the International Conference for the History of Cartography. A sucessful symposium on the history of cartography held as part of the 20th International Geographical Congress in London in 1964 persuaded a group of participants to establish an independent series of Conferences in co-operation with *Imago Mundi* but without any official parent body. The conference meets every two years, and is locally organized on an *ad hoc* basis. Its first meeting was in London in 1967, and it visited Brussels in 1969, Edinburgh in 1971, Warsaw in 1973, Greenwich in 1975 and Washington D.C. in 1977, to be followed by West Berlin in 1979. A selection of papers read at the conferences is published in *Imago Mundi*.

The founding of the Hermon Dunlap Smith Center at the Newberry Library, Chicago, and of the Kenneth Nebenzahl Jr. Lectures in the History of Cartography held at the library, provide another focus for map curators and others of kindred interest to meet. David Woodward, Director of the Center, describes its history and activities in his contribution to this volume.

Map librarians join in the work of these national and international organizations without always being the largest group of participants. This is especially true of the profession's relationship to cartographic organizations. Maps are produced and used by a wide range of official and commercial agencies, and liaison between map librarians and such bodies is mutually beneficial. An example of useful liaison of this kind is that between the National Cartographic Information Center established by the U.S. Geological Survey in 1974 and the SLA Geography and Map Libraries Division. In this way map librarians now move in more diverse spheres than their predecessors before the second World War and this widening of horizons adds greatly to the interest and efficiency of professional work.

At the same time the entrant into the profession still does not arrive with a full vocational training behind him. In Great Britain, for example, the only library course offered is the C203 option in Part II of the Library Association qualifying examination for Associateship, "Bibliography and librarianship of geography". As Muriel Lock pointed out in the chapter on "Map Librarianship" in *Modern maps and atlases* (1969), p.467, students in the time available can obtain no more than an introduction to the study of early and modern maps and atlases. Diploma courses in geography and cartography at colleges of technology now also provide a training, one which is especially suitable for map curators in university departments of geography. How far the training for map librarians in this and other countries answers the needs will be revealed by Lothar Zögner's investigation on behalf of the Geography and Map Libraries Section of IFLA. The proposed handbook for map curators would also fill a gap, for there are still few general textbooks on map librarianship. The Drexel Library Quarterly's issue on map librarianship, edited by J.B. Post and published in 1973, produced a useful set of articles, while Roman Drazniowsky's *Map Librarianship: Readings* (1975), comprising a collection of 48 articles which have appeared in journals, ranks as the first textbook on the subject. The second, Harold Nichols' *Map Librarianship* (1976) treats the subject from a British point of view. The literature is therefore not large outside the periodical field, but this is partly counterbalanced by the many works on cartography which are being added to the map librarian's reference shelves and to which he is himself contributing. The *1976 Bookman's Yearbook: Part 1. Maps and Atlases,* printing the proceedings of the Preconference of the Rare Books and Manuscripts Section, Association of College and Research Libraries, comprises papers given by leading map librarians and historians of cartography in North America. It well illustrates the benefit of liaison between dealers and publishers, librarians and collectors.

Whatever his background, the new recruit to map librarianship in the late 1970s can be assured of one thing; he enters a well-knit international community of map librarians and of others concerned in one way or another with maps. Many interests and disciplines are united when people come together in the pursuit of cartographic enlightenment and in the service of an increasingly map-minded public. Many friendships have been made to bind together this community of kindred spirits. This volume itself testifies to the personal friendships which have enriched these recent years in which map librarianship has come of age.

TRAINING FOR MAP LIBRARIANSHIP
Working Group in the IFLA Section Geography and Map Libraries

By Lothar Zögner

The training of map librarians is one of the questions to which Walter W. Ristow has devoted several articles during the past years (see Ristow 1969, 1976). It was a matter of great importance to him to draw attention to the present state of affairs and to encourage further developments. For this reason it seems particularly appropriate to offer a general discussion of this subject within the framework of this volume in honour of Walter W. Ristow.

The gratifyingly positive development of map librarianship in the last few decades has resulted in increasing awareness of the basic problem of training suitable, skilled, and competent staff. The IFLA Subsection "Geography and Map Libraries", founded in 1969, soon turned its attention to this problem. It was clear that here was a forum in which to discuss the education and training of map librarians. On the initiative of the then chairman of subsection Ib R. Kejlbo, a meeting took place in West Berlin in November 1973, with the Map Department of the Staatsbibliothek Preußischer Kulturbesitz as host. Colleagues from Denmark, Sweden, Finland and Norway reported on the present situation of training in map librarianship and discussed the theoretical and practical principles of a training scheme. In the subsection's meeting at the 1974 IFLA Conference in Washington D.C. I had the honour of reporting on the results of the West Berlin meeting. The Subsection decided to found a Working Group "Training of Map Librarians", the starting-point for further discussions leading to the development of general guidelines. The following are members of the Working Group: L. Zögner (FRG) Chairman, P.K. Clark (UK), I. Kodes (USSR), E.H. van de Waal (Netherlands), J. Winearls (Canada), J. Wolter (USA).

On the occasion of the IFLA conferences at Oslo in 1975 and Brussels in 1977, special sessions of the Geography and Map Libraries subsection (now section) met to discuss the tasks of the Working Group and the problems arising from educational needs in map librarianship. Members also reported on the current situation of educational activities in their respective countries. The aim of the Working Group is to provide a forum of information and coordination and to encourage national developments. It has the following terms of reference:

a) to investigate the current situation in respect of the training of map librarians and to make recommendations;

b) to prepare practical guidelines for map librarians;

c) to promote and organize seminars.

First of all, the Working Group plans to organize a pilot project in the form of a seminar, which will probably be held at Utrecht (Netherlands) in 1981. This is intended specially for participants from developing countries. If it is successful, further seminars will follow, either at the same place, in other countries or in the developing countries themselves. We are thus addressing ourselves to a particular target group, which we consider extremely important for a variety of reasons. Remembering the considerable difficulties facing the old-established map libraries in our countries in the past, we now have the chance to help the young developing countries to provide from the very beginning an appropriate place for map libraries in their library and information systems (see Wallis 1976). On the other hand, however, the developing countries also lack the advantages of a long and gradually established collection as a point of reference and a model. Here, the necessary guidance must be available as regards the setting-up, the equipping and the appropriate development of a map library. Whereas in our countries many institutions can fall back on their own more or less adequate map collections, in the young developing countries many public authorities and other institutions are dependent on the services of a map library, which must be correspondingly well prepared and equipped. Unsureness when dealing with cartographic material will make intensive assistance from map librarians necessary in the selection and use of cartographic material and in the interpretation of maps. The new map librarians must also become familiar with the special physical problems connected with climatic conditions. Here I refer the reader to the activities of the IFLA Section on Conservation in formulating recommendations for proper preservation in tropical and desert climates. These and other considerations show the vital importance of specialized staff, particularly in developing countries. The framework of the abovementioned seminar should thus convey both basic general information and references to specific problems.

However, a seminar lasting several days ought not to be the sole means of equipping future map librarians for their jobs in developing countries. Supplementary practical courses are necessary. Everyone who is involved in map library work knows how important day-to-day experience is. For this reason, beginners should be offered the chance of gaining practical experience in the daily work of map libraries by means of extended practical courses in appropriate institutions. We plan to compile a list of suitable map libraries willing to accept such trainees. Detailed information about languages, size and structure of the collections, catalogue and classification systems, technical facilities etc. ought to help anyone interested to find and contact the map library best suited to his needs.

To supplement his practical training the new map librarian also needs theoretical instruction in the form of a practical guide and working manual. The Working Group plans to compile such a manual, to be used at first in our seminar and then - with any necessary revisions - to be published as a "Manual for map librarianship with special reference to developing countries". This means that the manual will be directed at a specific target group. However, we believe that it will also serve as a basis and stimulus for further publications in this field. For, on the whole, there is still a need for suitable literature for the student of map librarianship, although positive developments have taken place in recent years and there seems to be a general endeavour to satisfy this long-felt need. A series of handbooks on map librarianship has been published, mainly as a result of practical experience and work during training courses: for example, R. Drazniowsky's "Readings" from his lectures at Columbia University's School of Library Service; the Graduate School of Library Science of Drexel University has published an issue on map librarianship; and H. Nichols, author of a handbook on map librarianship, gives lectures at the Department of Library and Information Studies at Loughborough University of Technology (UK).

As well as general progress in map librarianship a positive development in education and training possibilities is clearly visible. However, owing to the different conditions and importance of map librarianship in individual countries, there are considerable differences between available training schemes. Without doubt, the Anglo-American countries offer the most possibilities. An up-to-date account of the situation in North America, Great Britain and Australia may be found in Cornelius (1978). In the USA and Canada, five Library Schools have been offering regular specialized courses since about 1970 (see Ristow 1976) and there is a further range of very different training possibilities, some of which were described in cartographic and library journals. In most European countries, the training situation corresponds to the low number of completed map libraries and the resulting low demand for staff in these countries. Existing initiatives here frequently come from the important map libraries.

Today the map librarian is faced not just with constantly increasing production but also with a great variety of cartographic materials. He has to deal with new techniques in cartographic science as well as in the administration of the material. These new, increased and, in part, changed requirements demand a professional education and corresponding training standards. The training background of map librarians was and still is extremely varied. The integration of geography/cartography and librarian training has, however, already taken on concrete form. For example, in North America 20 per cent of the map librarians have a library science degree (Ristow 1976) - already 20 per cent, or only 20 percent, depending on one's point of view. But we are already facing the next obstacle: the impending conflict with documentation science. However, these new trends should not cause us to neglect the demand for a degree in cartography/geography as a basic requirement. It remains to be said that the qualitative and quantitative increase in library staff brings with it the possibility and the necessity of increased differentiation in the field of activity of a map library.

In order to establish a satisfactory training programme, we first need accurate studies of the map librarian's field of work and an analysis of demand. The Working Group will have to devote its attention to the questions arising in this connection. We need information on the type of staff required and what functions and tasks they will have to perform. We also need to know how far currently available training possibilities match the requirements and how far adjustments may be necessary.

Then we will be in a position to establish guidelines for a future training plan. The content of the training programmes must be based on the fact that we have a library with special material, special methods and special functions. This means that special knowledge is required for administrative and scientific purposes and a different specialized knowledge for the user and reference service (cf. Zögner 1975, 1976). The structure of training programmes should be based on and should take into consideration the differentiated field of a map librarian's work. Thus different training programmes must be developed in order to cover different conditions and different requirements:

1. continuing training for map librarians in order to keep them up to date with new developments and to fill in gaps in basic knowledge ;

2. specialization in map librarianship for librarians
   a) with knowledge of geography/cartography
   b) without knowledge of geography/cartography;

3. elementary introduction for librarians.

This last point will also be given special attention by the Working Group. We would like to encourage library schools to hold introductory courses on map librarianship for all librarians. Minimum information should be given on cartographic material and its service possibilities. This will certainly have a positive effect on general library work, for in the interest of the reference service every librarian should also be able to give information about cartographic material and to direct the user accordingly.

Cartographic material occupies an important place within the information and communication system of our modern society. The setting-up of efficient map libraries also requires us to concern ourselves with developing training possibilities for map librarians. In all countries, map libraries with their special material are faced with very similar tasks and problems. This is why the Working Group of the IFLA Geography and Map Libraries Section wants to promote and to influence the training of map librarians on the basis of international cooperation.

Literature on the training of map librarians

CORNELIUS, H.F.: Education for Map Librarians. In: *The Globe*, 9, 1978, p.36-57.

EHRENSVÄRD, Ulla: En kartbibliotekarie - vad är det? In: *Notiser från riksbibliotekarien*, 1974, p.22-27.

HAGEN, Carlos B.: Education and training in map librarianship. In: Special Libraries Association. Geography and Map Division. *Bulletin*. 77, 1969, p.3-7.
(Repr. in:) DRAZNIOWSKY, Roman, (comp.:) Map Librarianship, Readings. Metuchen, N.Y. 1974, p.413-423.

Institute Workshop on Map Librarianship. [Graduate School of Library Science, Drexel University, Philadelphia.] In: Special Libraries Association. Geography and Map Division. *Bulletin*. 64, 1966, p.16.

KIRALDI, Louis: Courses in map librarianship. In: *Special Libraries*. 61, 1970, p.496-500. [Courses at Western Michigan University, School of Librarianship.]

KOSACK, Hans-Peter: Die Ausbildung von Kartenbibliothekaren. In: *Kartensammlung und Kartendokumentation*. 9, 1971, p.117-124. [Report about the first special course in the Federal Republic of Germany at the Library School at Köln.]

LARSGAARD, Mary: Map Librarianship. An introduction. Littleton, Col. 1978. [Education for map librarianship. p.288-230.]

LOCK, C.B. Muriel: Modern maps and atlases. London 1969. [P. 467: Map Librarianship and education in Great Britain.]

Map Librarianship. A course offered at Columbia University School of Library Service. In: *Canadian Cartographer*, 1969, 2, p.165.

Map Librarianship. A course at School of Library Science, University of Toronto. In: *Canadian Cartographer*, 1970, 2, p.149.

Map Librarianship. Ed.: Jeremiah B. POST. - Philadelphia, Pa. 1973, 90 p. (Drexel Library Quarterly, 9, 4.)

Map Workshop for Librarians. Western Illinois University, 1973. Map collections [papers]. In: *Illinois Libraries*, 56, 1974, 5, p.341-430.

NICHOLS, Harold: Educating the map librarians. In: *Bulletin. Society of University Cartographers*, 11, 1977, p.23-26.

RAUCHLE, N.M.: Towards intellectual foundations of map librarian education. In: *Austr. Libr. J.* 24, 1975, p.94-99.

RISTOW, Walter W.: Map Librarianship. In: *Library Journal*, 92, 1967, p.3610-3614. [German version:] Das Kartenbibliothekarswesen. In: *Kartensammlung und Kartendokumentation*, 6, 1969, p.35-45.

RISTOW, Walter W.: Map library education in the United States and Canada. Paper pres. at the meeting of the Geography and Map Libraries Subsection, IFLA, Oslo 1975. In: *Inspel*, 11, 1976, 3, p.95-99.

SLEURINK, H.: Een eerste poging tot een opleiding tot kaartbeheerder. In: K.N.A.G. *Geografisch Tijdschrift*, 8, 1974, p.462-463.

THATCHER, Edward P.: Toward education in map librarianship. In: Western Assoc. of Map Libr., *Information Bulletin*. 3, 1972, p.2-3.

WALLIS, Helen: The training of map curators in Great Britain Paper pres. at the meeting of the Geography and Map Libraries Subsection, IFLA, Washington, D.C. 1974. In: *Inspel*, 10, 1975, p.28-32.

WALLIS, Helen; Peter CLARK: Education and training of map librarians. Special needs of developing countries. Paper pres. at the meeting of the Geography and Map Libraries Subsection, IFLA, Oslo 1975. In: *Inspel*, 11, 1976, 3, p.100-102.

WILKES, Margaret; David FERRO: Practical course in map curatorship. Edingurgh, March 31 to April 3, 1976. In: *Cartographic Journal*, 1976, p.106.

WILKES, Margaret; David FERRO: Map Librarianship-training in Scotland. In: SLA. Geogr. and Map Division. *Bulletin*, no. 137, 1977, p.201-205.

WINEARLS, Joan: Report on map librarianship course at Toronto. In: Western Assoc. of Map Libr., *Information Bulletin*. 3, 1972, p.5-7.

WISE, Donald A.: United States Dept. of Agriculture, Graduate School. In: SLA. Geogr. and Map Division. *Bulletin*, 99, 1975, p.39-42. [Course on map librarianship.]

WOLTER, John A.: A proposed three-week course for PAIGH map librarians. Paper pres. at the XIth consultation on cartography of the IXth General Assembly of the Pan American Institute of Geography and History. June 5, 1969. 18 p., 12 ill.

WOODS, Bill M.: Training for map librarianship. In: *Special Libraries*, 1952, p.87-88; p.102.

WOODS, Bill M.: Map Librarianship. In: SLA, Geography and Map Division. *Bulletin*, 23, 1956, p.9-12. [Proposal about the training of map librarians to the Subcommittee on Special Library Education of the Council of National Library Associations.]

WOODS, Bill M.: Map Librarianship. In: *Journal of Education for Librarianship*, 1966, p.9-12.

WOODS, Bill M.: A continuing need: education for map librarianship. In: SLA. Geogr. and Map Division. *Bulletin*, 81, 1970, p.28-29.

ZÖGNER, Lothar: Report of the West Berlin meeting of November 1973 and some further thoughts on the training of map librarians. Paper pres. at the meeting of the Geography and Map Libraries Subsection, IFLA. Washington, D.C. 1974. In: *Inspel*, 10, 1975, p.33-37.

ZÖGNER, Lothar: Training of map librarians - present situation and future programme. Paper pres. at the meeting of the Geography and Map Libraries Subsection, IFLA, Oslo 1975. In: *Inspel*, 11, 1976, p.103-108.

ZÖGNER, Lothar: Zur Ausbildung von Kartenbibliothekaren. In: *Zeitschrift für Bibliothekswesen und Bibliographie*, 23, 1976, 3, p.172-175.

THE AMERICAN GEOGRAPHICAL SOCIETY'S COLLECTION

By Roman Drazniowsky

The development and function of the Library and Map Collection of the American Geographical Society is better understood after one is acquainted with the background of the Society and the purpose for which it was established.

By 1850 New York had developed into a booming metropolis. The demand for information on subjects related to natural resources, trade, and exploration of recently discovered regions, especially in North America, accentuated the need for a geographical society in the city. The business and professional community of the city, following the example set in foreign countries, considered establishing such a society.[1] Preliminary efforts were made to explore the possibility of organizing an American geographical society, based in New York.

In September of 1851 a number of influential New Yorkers met and appointed a committee to draft a constitution for the new organization. On October 9th, at a public meeting held in John Disturnell's Geographical and Statistical Library at 179 Broadway, the constitution was adopted and Mr. Henry Grinnell was elected president of the American Geographical and Statistical Society. The Society adopted its present name in 1871.

The purpose of the Society and its future development were clearly formulated in Article 1 of the Constitution, and special emphasis was placed on the establishment of a library. The article reads, "The American Geographical and Statistical Society of New York is instituted for the collection and diffusion of geographical and statistical information."[2] In addition, the Society was to publish its own transactions, with maps and illustrations, and also to "possess a library with a collection of maps, charts, and instruments."[3] In order to incorporate the Society a draft of a charter was read at a business meeting on February 21, 1852. In April of the same year, both houses of the New York State Legislature voted to grant the charter. However, for undisclosed reasons, Governor Washington Hunt vetoed the bill.

The Council of the Society was not disappointed by the governor's veto - quite to the contrary - and rooms in the old building of New York University were rented. Numerous meetings were held and a collection of books and charts, donated to the Society as early as 1851, were housed at that location. The nucleus of the Library consisted of fifty volumes of the documents published by the New York State Legislature and a set of Maury's Wind and Current Charts and Pilot Charts.

In 1854 the Legislature again voted to grant the Society a charter, and this time the bill was approved by Governor Horatio Seymour.

The ever-increasing need for more information prompted the Council in 1855 to authorize the subscription to four foreign periodicals: Zeitschrift für allgemeine Erdkunde (Berlin), Bulletin de la Société de Géographie (Paris), Das Ausland (Stuttgart), and Nouvelles annales des voyages et des sciences géographiques (Paris).

Cooperation among geographical societies and other scientific societies throughout the world led to the exchange of their publications. The exchange program implemented by the Society worked so well that it was possible to acquire a considerable amount of library material on a very limited budget. The same holds true today. Exchange agreements, some of them lasting more than one hundred years, were arranged with the Royal Geographical Society in London, the Royal Society of Edinburgh, the Geographical Society of Bombay, and many others.

The growth of the collection demanded a professional staff. The first staff member, Dr. Edward R. Straznicky, was appointed to the post of Librarian in 1857. His first accomplishment was the compilation of a catalog of the Library's holdings.

In 1858 it became apparent that the collection, which included 3200 volumes of books and periodicals, 3150 pamphlets, and 500 maps, needed more space. Therefore, the Society and its collection were transferred to Clinton Hall, a remodeled opera house at 8th Street and Lafayette Place. Through the generosity of Peter Cooper the Society moved again in 1867 to larger, rent-free quarters at Cooper Union, where it remained until 1876. That year, through the efforts of President Charles P. Daly, a brownstone house was purchased at 11 West 29th Street. Once again, the increasing size of the collection had necessitated the move. In 1870 the collection of books, periodicals, maps, and charts numbered more than 10,000 items.

Treasures such as the Ortelius atlas of 1573 indicate the quality of the collection at that time; however, its real value was in the excellent classification system and easy retrieval method for needed material, characteristics which have remained to this day.[4)]

The demand for more information, brought about by the expanding amount of research done by the Society, increased the size of the collection. As a result, in 1902 the collection was divided into two separate units, the Library and the Map Department.[5)] The new position of Map Curator was created and Mr. Thomas Letts was appointed to catalog and arrange the map collection. The Society's first Map Curator soon demonstrated his professional ability by presenting a paper entitled "Maps: their Handling, Classification and Cataloging" at the Eighth International Geographical Congress, which was held in New York in 1904. His regional arrangement of maps and analytical cataloging of atlases have been modified but are still in use to this day.

In 1911 it was necessary for the Society to move again. Its new site was on Audubon Terrace, at Broadway and 156th Street, where it is now located. Mr. David Randall-MacIver, then the Society's Librarian, and Mr. Thomas Letts, the Map Curator, planned the relocation and arrangement of the material.

The Map Department was established on the third floor of the spacious new building. Special care was taken that the books and maps were properly stored. The entire stack area designated for book storage received new shelves, constructed by Snead & Co. Iron Works, Inc., located in Jersey City, New Jersey. This same company constructed shelves for many other libraries, among them the Library of Congress. Horizontal map cases, specially designed for the Society, were custom-made on the premises, and are still considered the best means of map storage.

The growth of the new unit, the Map Department, demanded more than one staff member. Therefore, shortly after the relocation of the Society in 1911, Mr. Leon Dominian joined the staff as Assistant Map Curator.[6)] Because of his linguistic ability, Mr.L.Dominian also performed additional services for the Society. Much of his time was given to editorial duties for the Society's Bulletin. This practice of utilizing the services of the Map Department staff for other duties within the Society was, unfortunately, repeated many times and had a detrimental effect on the operation of the Map Department. Since it was obvious that the inadequate staff could not provide required services, Mr. J.R. Wallace was hired in 1914 as Map Room Assistant.

Dr. Isaiah Bowman, the first Director of the Society, was appointed on July 1, 1915. He both appreciated and made considerable use of the Society's collection. He realized that a reputable research organization with an excellent collection on geography, history, economics, statistics, politics, and many other subjects, subjects so vital especially in time of war, had the potential to be of immense assistance to the United States government. As a result of his efforts an organization called the "Inquiry" was set up in the Society's building to study, select, and assemble material for the Peace Conference at Versailles. The extensive and diversified collection of the American Geographical Society contributed a great deal to the work carried out by the "Inquiry", especially in the cartographic field. The importance of the collection and its use was evident when members of the "Inquiry" sailed to France for the Peace Conference. They borrowed and took with them thousands of books and maps from the Society's collection.

After the war Mr. Bowman again demonstrated his ability to utilize the Society's collection. He started one of the most important cartographic projects ever undertaken by the Society, the compilation and publication of the *map of Hispanic America* on a scale of 1 : 1 000 000. It was possible to undertake the 25 year project (1920-1945) because of resources available in the Society's collection.

During the early years of Isaiah Bowman's directorship important decisions were made, not only in regard to the projects, but also in regard to staff. In 1917 a new assistant, Ena L. Yonge, was appointed to the Map Department. At that time no one could foresee that the newest and youngest appointee would develop the Society's map collection into one of the best of its kind. Shortly after her arrival Mr. Dominian resigned, and under the guidance of Mr. W.L.G. Joerg, then the Assistant Editor of the Society's Bulletin and extremely knowledgeable about maps, Miss Yonge began her map librarianship training. By 1927 her knowledge of maps and ability to operate the Map Department were superb, and the Director decided to appoint her as Map Curator. Through her tireless efforts she managed to greatly increase the size of the collection, always stressing that "maps and atlases do not appear on the shelves by magic". Between 1917 and 1943 approximately 2450 maps were added annually to the collection.[7)]

The second promising staff member was Dr. John K. Wright, who later became Director of the Society. In 1920 he was appointed to the prestigious post of Librarian. One of his first projects was the initiation of the now famous Research Catalogue. Dr. Wright, being a geographer as well as a librarian, felt that the conventional dictionary catalog was inadequate for geographical research and, as a result, he developed a hierarchically arranged catalog, based on the classification system utilized by the Berlin Geographical Society. Dr. Wright, in a letter to Miss Nordis Felland, successor to the post of Librarian, wrote in 1967, "the most important thing I ever did was to initiate the Research Catalogue".

The importance of the Research Catalogue for scientists, and researchers was evident when in 1938 *Current Geographical*

Publications, a monthly bibliography, was inaugurated. It included references to books, articles in periodicals, monographs, pamphlets, government documents, technical reports, maps and atlases received by the Library and processed according to the Research Catalogue rules.

The importance of the Library, and particularly the map collection, to projects of the American Geographical Society is evident from Dr. Bowman's memorandum:

February 15, 1924

To: Miss Wrigley
Yonge
Mr. Wright
Joerg
Platt
From: Mr. Bowman

The persons whose names stand at the head of this memorandum are constituted a Map Committee, with Mr. Wright as Chairman. The purpose of the Committee is to study the map collection of the Society, determine what steps should be taken for its improvement, and make recommendations from time to time. It is for the Committee to study the whole question, but I would make one specific comment now to illustrate the nature of the work which the Committee should undertake:

We have a large number of map sets in our collection. They are of various dated and of various degrees of usefulness to consultants and especially to members of the staff engaged upon research work. No one has made a critical selection of such sets from among those available. It should be one of the objects of the Committee to make such a selection.

To take another illustration: we have acquired perhaps a thousand atlases. When a new atlas comes upon the market, what test have we of its value in our series? Naturally, an atlas might be cheap as to money cost but expensive to us because we may have representative atlases of the same period. Again, a low priced atlas listed in a catalogue we might find to be of very great value to us because of an entire absence of atlases of that period in our collection.

A third illustration related to the catalogue. Does the

catalogue work? Is the present system a desirable one? There are many tests to be applied by different people with different objects in mind. Have we a complete set of the topographic sheets of the U.S. Geological Survey? In a number of instances certain sheets have been missing. Under what conditions are they missing?

The Committee ought to have at least one meeting a year and the work of the Committee ought to be divided among its various members, and there should be a formal report once a year not later than December 31 so that significant items may be incorporated in the Annual Report of the Chairman of the Council.

The response was good and on February 19, 1924, Dr. Wright sent a memorandum to Committee members:

To : Miss Wrigley
Miss Yonge
Mr. Joerg
Mr. Platt
From: Mr. Wright

Referring to Dr. Bowman's memo of February 15 constituting a map Committee, I should like to make the following suggestions.

1. *All members of the Committee at their convenience to submit suggestions to me in writing on any of the topics relating to the map collection.*

2. *The following members to make a special point of giving advice on the following topics:*

   a. *Miss Yonge on ordering, care, and classification of maps and atlases;*

   b. *Mr. Joerg on map sets and modern atlases;*

   c. *Mr. Platt on maps of Latin America;*

   d. *Myself on old maps, old atlases, and on the catalogue.*

3. *The Committee to prepare a brief report on special desiderata before May 1, 1924, in order that it may be in Dr. Bowman's hands before the summer vacation.*

4. *The Committee to hold a meeting in November or December for discussion of its formal annual report.*

February 19, 1924. *JKW*

The growth of the Library and especially the map collection was remarkable.

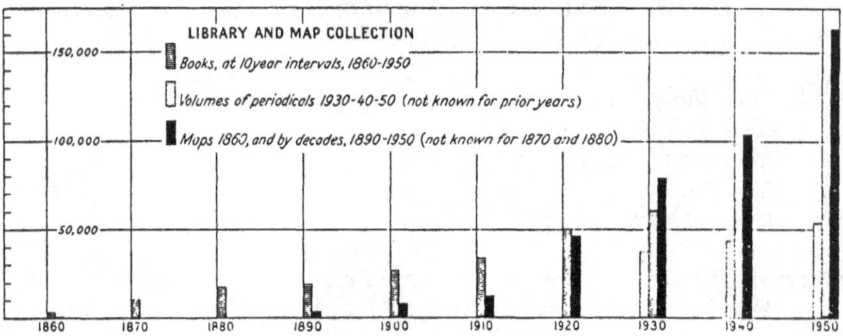

However, the financial growth of the Society was not so remarkable and the Great Depression did not help to improve a worsening financial situation.

With the resignation of Dr. Bowman from the Director's post in 1935, the collection was consistantly faced with budgetary difficulties, although it continued to grow, due to the efforts of a dedicated staff. In order to redefine the collection's policy and to establish priorities for urgent programs, a conference was held in the Society's building on November 4, 1948. Notable professionals were invited to participate: Mr. Nathaniel O. Abelson - Map Librarian, United Nations, Dr. Charles W. David - Director of Libraries,

University of Pennsylvania, Dr. Stephen B. Jones - Professor of Geography, Yale University, Mrs. Clara LeGear - Map Division, Library of Congress, Mr. Gerald McDonald - Division of American History and Geneology, New York Public Library, Dr. Carl White - Director of Libraries, Columbia University. The following staff members of the American Geographical Society participated: Dr. Richard U. Light, Dr. John K. Wright, Miss Nordis Felland, Miss Ena L. Yonge. The importance is reflected in the agenda.

AGENDA

The conference is asked to consider the following undertakings in terms of their scientific and scholarly value, feasibility, costs, etc., and to make recommendations as to the priorities that should be assigned them by the Society in seeking support for their execution:

First priority: *urgently needed*
Second priority: *desirable, but less urgent*
Third priority: *to be dropped or indefinitely postponed*

A. *Library and Map Department*

1) *Cataloguing uncatalogued books*
2) *Cataloguing uncatalogued maps*
3) *Catching up with binding, map mounting, purchase of new equipment, etc.*
4) *Future maintenance of collections and catalogues in first rate condition*

B. *Bibliographical*

1) *Publication of a complete catalogue of the collections*
2) *Improvement of bibliographical services now being rendered through "Current Geographical Publications" by:*

a) Publication of cumulated edition for years 1938-1947
b) " " " index " " " "
c) Enlargement and improvement of current numbers

3) *Publication of special bibliographical works:*

   a) *With emphasis on maps and atlases*
   b) *Check-list of geographical periodicals*
   c) *Progress reports on geographical research*

4) *Union catalogues of geographical literature*

5) *Punch-card system of cataloguing for bibliographical service*

The need for such a conference had been evident for a long time. Difficulties in the operation of the collection were not only of a financial nature. Due to the Second World War the unexpected increase of donated maps and requests for information brought feverish activity to the Map Department. The collection was used by more than forty government agencies to secure information not available elsewhere. Microfilms of the 113 000 cards of the Research Catalogue were distributed by the State Department to a number of government agencies. Due to the increase in activity, which continued after the war, Mrs. Mary E. Harzell was hired as a full-time assistant in the Map Department in 1944.

Designation of the Society as a map depository by the Army Map Service caused the map collection to increase rapidly in size. The Society received approximately 84 000 maps in six years, more maps than it had acquired in its previous eight decades of existence.[8)] To process that amount of material an additional assistant, Mrs. Georgia Bergnes, was employed in 1946.

In May of 1962 Dr. Charles B. Hitchcock, Director of the Society, offered to the author the position of Map Curator. At the end of 1962 Miss Yonge retired after 46 years of dedicated service to the Society.

Appointment to the Map Curatorship was a formidable challenge. The financial situation at the American Geographical Society was not promising, yet demands were made by the administration to maintain the growth of the collection and to provide services to the staff of the Society and other patrons. To achieve that goal it was absolutely necessary to develop a careful plan for operation of the collection and to establish priorities for urgent projects. Previous recommendations were very helpful in establishing new policies. It was remarkable that some problems could be traced as far back as 1924.

One of the most urgent projects was the preparation of more comprehensive cataloging rules. By 1964 new *Cataloging and Filing Rules* were published and implemented. Major changes were made in defining more accurately regional and topical entries. Special attention was directed toward the expansion of analytical cataloging of atlases, particularly thematical atlases. As a general rule, the following catalog entries were prepared for maps and atlases:

1. Geographical area or political unit.
2. Topic represented on the map.
3. Publishing authority, cartographer, or editor.

The catalog card provided information on:

1. Area.
2. Topic.
3. Date of the situation represented on the map.
4. Scale.
5. Projection.
6. Insets.
7. Cartographer or editor.
8. Publishing authority, place of publication, and date.
9. Revision and/or edition.
10. Brief description in regard to language, colors of the map, and topics.

The unique task of cataloging maps in books and periodicals was considerably intensified. To succeed in maintaining the quality of the cataloging and the high level of production, a new full-time cataloger was added to the staff of the Map Department. In 1968 this increase in cataloging paid off. In that year ten volumes of an important cartobibliography, the *Index to Maps in Books and Periodicals*, based on the Society's Map Catalog, were published by G.K. Hall. Since that time two supplements have also been compiled and published.

It was evident that with the number of maps distributed by the Army Map Service depository program diminishing, efforts must be made to maintain the acquisition level of maps by other means. In the past the exchange program was one of the more important acquisition sources. It was decided to expand and to intensify that program.

In order to encourage map publishers to send maps and atlases for the American Geographical Society's collection, the material was included in Current Geographical Publications. The following statistics reflect the importance and success of the exchange program and solicitation of free material.

Material Received

Year	Atlases			Globes			Maps			Reference		
	G	P	T	G	P	T	G	P	T	G	P	T
1963	67	10	77	2	-	2	5455	149	5606	94	5	99
1964	72	25	97	-	-	-	6640	1000	7640	194	11	205
1965	68	21	89	-	-	-	5467	820	6287	92	13	105
1966	121	28	149	3	-	3	6976	783	7759	49	4	53
1967	125	27	152	-	-	-	7430	623	8053	62	2	64
1968	141	30	171	-	-	-	8252	836	9088	27	7	34
1969	80	9	89	1	-	1	6730	230	6960	28	8	36
1970	87	46	133	8	4	12	5721	514	6235	55	12	67
1971	70	31	101	-	-	-	5572	317	5889	19	2	21
1972	72	34	106	-	-	-	4771	527	5298	26	5	31
1973	87	30	117	1	-	1	5041	313	5354	30	14	44
1974	98	6	104	1	-	1	7289	18	7307	13	5	18
1975	89	7	96	1	-	1	4760	7	4767	49	8	57

Note: The letters used in column headings stand for the following: G - Gift; P - Purchase; T - Total

Due to extensive political changes and the addition of many topical entries, a revised and enlarged edition of the "Cataloging and Filing Rules" was published in 1969.

Because of improved cataloging procedures and an excellent retrieval system it was possible to provide more reference services to the Society's researchers and outside patrons.

		Consultants			
Year	Staff	Visitors	Others	Telephone	Correspondence
1963	290	348	-	473	83
1964	318	663	-	727	82
1965	507	592	-	790	151
1966	651	699	237	941	192
1967	841	667	242	911	222
1968	768	475	382	780	207
1969	509	339	323	473	184
1970	673	416	6 classes (219 students)	644	275
1971	528	434	-	502	249
1972	575	428	2 classes ((72 students)	510	287
1973	718	457	1 class (35 students)	522	273
1974	826	746	5 classes	647	293
1975	604	71	7 classes	377	189
				xerox	479

Unfortunately, a number of programs could not be implemented because the necessary funds were not available. One such program was the restoration and repair of rare maps and globes. Badly needed filing equipment could not be purchased. The installation of environmental control equipment in the storage area had to be postponed. However, the collection continued to develop more rapidly than other units of the Society, with potentially fatal results for it as well as for the Society. Regrettably, recommendations made in 1956 were never implemented.

Conclusion

The heart of the Society's financial program, as of its many activities, is the library and map collection. Some way must be found to assure the continuation of these essential collections. Other activities of the Society can be adapted so that they will bring in the revenues needed to carry them on. However, the collections by their very nature cannot be revenue producing.

The endowment of the Society must be increased by $700 000 to be vested in a Library Endowment Fund. Only thus can the future of these collections, so vital to the future of geography in the United States, be assured.

As a result of the strained budget of the American Geographical Society, in December of 1975 the collection staff was drastically reduced. At that time no one expected that it would be possible to provide services and to publish Current Geographical Publications with the remaining staff. But miracles happen; the collection is open, limited services are provided, and Current Geographical Publications is published. However, time is running out and the Society alone can no longer maintain the collection. Desperate efforts to save the collection were made by the staff but it is evident that this is impossible. The only hope to preserve the collection is to amalgamate it with another institution which can cover its operating costs. In 1976 negotiations to that effect were conducted and an agreement reached with the University of Wisconsin in Milwaukee. Under the terms, the University agreed to accept the collection, preserve it as a unit, continue present cataloging methods and publication of Current Geographical Publications, provide adequate housing, maintain acquisition levels, support it with a full-time staff of twelve, and, most important, restore rare maps, atlases, and books.

Unfortunately, legal complications have prevented the implementation of this agreement up till now. It is sad to witness the deterioration and possibly complete destruction of one of the most well-organized special collections because of legal difficulties.

The following statistical compilation gives some indication of the development and rapid growth of the collection.

Status of the Collection

Year	Atlases	Globes	Maps
1963	4,039	-	282,316
1964	4,137	47	289,826
1965	4,211	47	294,537
1966	4,389	47	301,895
1967	4,498	50	309,407
1968	4,661	50	318,393
1969	4,734	50	319,617
1970	4,844	55	323,169
1971	4,961	55	328,671
1972	5,076	55	332,066
1973	5,191	56	336,241
1974	5,276	57	342,776
1975	5,372	58	347,543

Based on the previous history, excellence, and prestige of the collection, it is obvious that the geography profession as well as the scientific community must make efforts to preserve and to continue to develop this unique collection.

NOTES

1) *Geographical societies were successfully established in Paris (1821), Berlin (1828), London (1830), Mexico City (1833), and Moscow (1845).*

2) WRIGHT, JOHN KIRTLAND: *Geography in the Making: the American Geographical Society, 1851-1951. New York: American Geographical Society, 1952, p.18.*

3) *Ibid., p.19.*

4) *Journal of the American Geographical Society, 2, 1870, no. 2. p.III-IV.*

5) WRIGHT, *p.XVI.*

6) Ibid., *p.141.*

7) *Ibid., p.360.*

8) *Ibid.*

THE HERMON DUNLAP SMITH CENTER FOR THE HISTORY OF CARTOGRAPHY

A Review of its Early Development

By David Woodward

Background

The collections of maps, atlases, and cartographic monographs and bibliographies in The Newberry Library have long been considered among its greatest strengths. While a small core of such items was to be found in the library since its establishment, the real beginning can be traced to Edward E. Ayer's interest in the discovery and exploration of the Americas, an interest that figured prominently in the forming of his great collection devoted to American Indians and their interaction with Western civilization. The Ayer Collection, given to the Library in 1911, regularly augmented by gifts from Mr Ayer until his death in 1927 and since then from his endowment, has such highlights as the Henry Stevens Collection of editions of Ptolemy's *Geographia* (including four manuscript versions), eight portolan atlases, thirteen portolan charts, and a very strong representative collection of maps, atlases and geographical works published before 1800.

In addition to the Edward E. Ayer cartographical materials, the Newberry has many more maps and atlases in its General Collection and in special collections, notably Everett D. Graff's on the American West and William B. Greenlee's world-famous gathering of Portuguese materials before 1830.

The planning for the Hermon Dunlap Smith Center for the History of Cartography was part of a wideranging policy to promote rigorously the use of the collections, a policy conceived and initiated by the present President and Librarian, Lawrence W. Towner, who, since his appointment in 1962, recognized the extent and importance of the existing collections for cartography. He perceived the need for developing the field of the history of cartography as an important branch of the humanities, and took several crucial steps toward the founding of the Center.

Much of the interest in plans for the new center was sustained by the Kenneth Nebenzahl, Jr. Lectures in the History of Cartography, founded in 1964 by Mr. and Mrs. Kenneth Nebenzahl in honor of their son, Kenneth Nebenzahl, Jr. (1954-1971). R.A. Skelton, then Superintendent of the Map Room of the British Museum, delivered the first series of lectures in 1966 on the history of the study and collecting of early maps. During his stay at the Library, Mr. Skelton prepared two reports, one summerizing the strengths and weaknesses of the map collections, the other recommending ways in which these collections could serve as a basis for a center for the study of the history of cartography. He stressed in particular the value of the location of The Newberry Library in what he called a "golden crescent" of rich cartographic and geographical resources in the Middle West: The James Ford Bell Library, the William L. Clements Library, the Lilly Library; eleven major state and private universities represented by the Committee on Institutional Cooperation; several major state historical societies and their libraries; and a score of other important repositories of cartographic materials.

Spurred on by this encouraging report, Mr. Towner acquired two major private map collections for the Library: the Franco Novacco collection of sixteenth-century Italian printed maps in 1967, and the Johann Gabriel Sack Collection of eighteenth-century European maps a year later.

Two more building blocks were necessary. Despite the strong cartographic collections of the Library, no full-time position as map curator or historian of cartography had ever been established. In 1969, David Woodward was appointed Fellow in the History of Cartography and Consultant to the Library. His task was to draft plans for rationalizing the Newberry's cartographic materials with the help of the previous reports by R.A. Skelton, and to make proposals for the establishment of a center for the history of cartography that could bring the Library's collections to the notice of potential users. Clearly, one of his first tasks was to become familiar with the Newberry's maps (many of them uncatalogued), and to consolidate some thirty-eight discrete physical locations of maps and atlases in the building into a manageable space. This problem was severe enough to warrant the establishment of a full-time curatorial position shortly after the establishment of the Center. Robert W. Karrow, Jr. was appointed as map cataloguer in 1971 and continued in that capacity for the Center until 1975 when he was named curator of maps. At the same time, the curatorial functions were transferred to the Department of Library Services.

A further development that focussed attention on the Library as a center for cartographic study, was the attraction of the Atlas of Early American History Project to the Newberry in late 1970. This project, sponsored by the Institute for Early American History and Culture and The Newberry Library, brought six historians and two cartographers during its five-year duration to the Library to work on a major historical atlas using among its varied sources the eighteen-century maps and atlases in the Library.

On March 15, 1971, the day of their annual meeting, the Trustees announced that a capital gift, the largest such gift ever received by the Library, has been pledged for the purpose of establishing a Center for the History of Cartography. This enabled the funding of a permanent position, and David Woodward was appointed as coordinator of the Center, a position he held concurrently with that of curator of maps in the Department of Special Collections, effective 1 April 1971.

The Hermon Dunlap Smith Center for the History of Cartography was formally established on 1 November 1972. At the dedication ceremony, the Center was named for the donor of the endowment, Mr. Hermon Dunlap Smith, then President of the Board of Trustees of the Newberry Library. The Center is the first privately endowed institute for research, education, and publication in the history of cartography, and its development over the last five years has demonstrated its potential for influencing the growth of this young subject on an international scale.

Purpose

The purpose of the Center has been to promote the study of the history of cartography at The Newberry Library. Within that seemingly simple statement there are important and involved implications. In the first place, since directors of research centers at the Library are given broad latitude, the specific interpretation of the goals of the Center for the history of cartography has depended in the last analysis on the leanings of its director. The director has accepted R.A. Skelton's distinction between the history of cartography and historical cartography. By "historical cartography" Skelton ment the reconstruction of past geography by cartographic means; in other words, the drawing of modern maps of historical landscapes and events, as exemplified in the *Atlas of Early American History* and the *Atlas of Great Lakes Indian History*. He thus objected vehemently to the first series of Nebenzahl lectures bearing the words "historical cartography" in their title. He felt that the scope of these lectures should be far broader: to describe and explain the development of mapmaking as an art, science, and technology in the context of other human efforts in cultural civilization, and it was this field that he felt should be described as "the history of cartography".

While the center purpose of the Center is to promote the history of cartography as defined by Skelton, its director recognizes that there are several ancillary fields that have traditionally been associated with the history of cartography through similar subject matter and approach. Among these may be cited the history of geographical discovery and exploration; historical geography; the history of geography; and, as noted in Skelton's distinction above, historical cartography. It is certainly not the intention of the director to exclude in dogmatic fashion the pursuit of such studies at the Hermon Dunlap Smith Center, but it must be stated, that in the face of limited resources, they must be considered as ancillary to the Center's main function.

Also central to the Center's purpose is the support of work pursued by scholars while in residence at the Newberry Library. Questions occasionally arise whether the Center can provide fellowship for scholars to study elsewhere, but this is contrary to the policy of the Center for two reasons: it is not financially feasible and it would not serve to develop the Center's role as part of the Newberry Library's community of scholars from different disciplines. The Center does help to support the wider cause by subsidizing in a modest way the publication of important work in the field, such as *Imago Mundi* and the recent *Map Making to 1900: An historical glossary of cartographic innovations* compiled by the Working Group on the History of Cartography of the International Cartographic Association, now a Commission.

Functions

It is convenient to describe the functions of the Center under three broad categories: education, publication, and research. The first part describes the teaching activities of the staff of the Center and the various courses of instruction; the section on publication outlines the ways in which the Center sponsors and subsidizes the publication of research in the history of cartography, and the last section describes the research projects, fellowship, and research associateships at the Center.

Education

Since the field of the history of cartography is in its early stages of development, the Center has regarded education to be a strong component of its activities. It has catered to various levels of students, both graduate and undergraduate from Northwestern University, the University of Chicago and other universities in Chicago. In addition, it has responded directly to the lay public in Chicago and the surrounding area, notably map collectors and map enthusiasts, and the widely spread group of scholars and amateurs in the history of cartography gathered here every two or three years for the Kenneth Nebenzahl, Jr., Lectures in the History of Cartography. Most of the efforts have been expended towards education at the local level, both academic and non-academic.

The evening course, "The History of Mapmaking," believed to be one of the first of its type, has now been taught six times at the Library by the Center's director. Its purpose is to provide a general introductory survey of the history of cartography. The structure of the course has been broadly chronological, with six evening sessions on the following topics: the Cartography of Non-literate Peoples, the Classical Period, the Medieval Period, the Renaissance, the Age of Atlases, the Enlightenment, and the Nineteenth Century. As the course developed, two more sessions were added in response to popular demand: the principles and simple techniques of map conservation, and the whys and wherefores of map collecting. The participants of these courses have come from widely different vocational backgrounds; about half have been map collectors, (many from the legal and medical professions), while the remainder have been either professionally involved with cartography (for example, as teachers, draftsmen, map editors, or map librarians), or simply interested in maps as a diversion from their daily tasks.

From these evening courses there arose one of the first formally organized societies of map collectors and enthusiasts, the Chicago Map Society, founded in January 1975. The society was created in direct response to the wishes of several participants in the evening courses to continue their studies in the subject in more depth. The Society offers monthly meetings, visits to private collections, exhibitions of members' collections,sponsorship of publications (such as the *Directory of Dealers in Antique Maps*), and the arranging of social events to allow fellow map collectors to meet and compare notes. The society has now grown to over a hundred members, and has apparently influenced the formation of several other similar societies: the Wisconsin Map Society, the Illinois-Missouri Map Society, the Michican Map Society,and, most recently, a group in New York City.

The Center's associations with local universities began in summer 1971 when the Director led an eight-week summer course/ seminar in the history of cartography under the auspices of the Department of Geography, Northwestern University, and sanctioned by the Committee on Institutional Cooperation of eleven midwestern universities. Attempts to continue this course in the summers of 1972 and 1973 failed through inadequate enrollment. A more successful association with the University of Chicago began in January 1975, with the Director leading a seminar in the history of cartography aimed at students in geography, history, and library science. This arrangement, through the Department of Geography, has since continued on a yearly basis as part of that department's regular program.

One of the problems faced by the Center is shared by more institutions of higher learning than would care to admit it: the difficulty of fitting a subject largely dedicated to pure research into the increasingly practical programs of study that are being developed, out of necessity, by student demand. The spirally costs of tuition and the higher competition for employment have led students to seek those courses that relate directly and practically to their chosen field of study, and to reject the marginal course or so-called *elective*, however beguiling the subject matter, on the grounds that it is an expensive luxury.

In response to this, the Center is developing educational programs that combine traditional classroom teaching with practical experience in map curatorship. One such program is designed on the internship principle for undergraduates in their junior or senior year who intend to follow a career in librarianship. The combination of research centers and an outstanding research library make the Newberry a unique laboratory for this kind of learning experience.

P u b l i c a t i o n s

The aim of the publishing program of the Hermon Dunlap Smith Center has been to attract manuscripts of major significance in the history of cartography, and to maintain a high quality of publication at all levels, from copy editing stage through design matters. In trying to achieve these goals, a close relationship with the University of Chicago Press has been established. The direct contact with the many departments of the Press and its convenient proximity have expedited the many route matters that arise in the process of publication. Three series of publications series have been sponsored by the Center: Studies in the History of Discoveries, the Kenneth Nebenzahl, Jr., Lectures in the History of Cartography, and the Center's own publication series.

Studies in the History of Discoveries

Studies in the History of Discoveries is the monograph series of the Society for the History of Discoveries, a small society with an international membership founded in 1961 by a number of scholars in different fields with a common interest in the history of geographical exploration and discovery. In 1963, the Trustees of the Library pledged sufficient subsidy to publish five books in the *Studies in the History of Discoveries* series over a period of ten years.

The first book to be published in the series, the *Proceedings of the Vinland Map Conference* (1971) edited by Wilcomb E. Washburn, presented the varied views of scholars in linguistics, church history, Scandinavian studies, the history of cartography, and paleography concerning the authenticity of the Vinland map. These papers, originally delivered at a symposium at the Smithsonian Institution in 1966, provide a valuable volume of commentary on *The Vinland Map and Tartar Relation* (New Haven: Yale University Press, 1965).

Although it was the first manuscript accepted for the series, Ursula Lamb's *A Navigator's Universe* did not appear until 1972. This work is a scholarly edition of a manuscript manual of cosmography by Pedro de Medina, and reproduces his 1538 manuscript in the Bodleian Library, Oxford. The treatise, known as the Libro de Cosmographîa, is arranged as a dialogue in which an educated laymen and a pilot puts eighty-two questions to a cosmographer, who answers them all, revealing the pre-Copernican cosmographical concepts with remarkable clarity.

Also published in the series in 1972 was T. Bentley Duncan's *Atlantic Islands,* a history of seventeenth-century trade and navigation in the Portuguese islands of the Azores, Madeira, and the Cape Verdes. This book had already been accepted by The University of Chicago Press, but since its subject matter was appropriate to the series, it appeared as the third title.

David B. Quinn's *The Last Voyage of Thomas Cavendish* is a fine example of careful and complete editing of a manuscript by the foremost scholar in this area of Elizabethan travel narratives. The manuscript is an account of Cavendish's unsuccessful 1591-92 voyage in which one of the five ships was lost and in which Cavendish apparently died at sea on the flagship *Leicester*. His journal, in addition to being of great historical interest, also describes his psychological decline towards the end of the voyage, and his systematic preparations for his own death. In addition to the textual importance of this book, it is a fine example of scholarly bookmaking: it was recognized by the American Institute of Graphic Arts as one of the fifty best designed books of 1975.

The last volume in the series was *The Discovery of South America and the Andalusian Voyages* by Louis-André Vigneras, a monograph on the subject of eleven so-called minor voyages to the north-east coast of South America that followed in the wake of Columbus, based largely on previously unpublished archival sources in Seville, where most of the voyages embarked.

Nebenzahl Lectures

The Kenneth Nebenzahl, Jr. Lectures in the History of Cartography are now well established. They not only fulfill a useful function in bringing together at the Newberry Library the lay and scholarly community of persons interested in the history of cartography; they have generated significant additions to its literature.

The format of the lectures has changed since their inception. With the first and second series, in 1966 and 1970, the lecturers were chosen for their widely known competence in the field, and were asked to speak on a topic of their choice.

The first lecturer, R.A. Skelton, delivered four lectures on the history of the study and collecting of early maps. The published version of these lectures provides an excellent historiography. Following a brief background history of cartography is an essay describing the many factors that have contributed to the disappearance of maps from the historical record. In the third and fourth essays, respectively, Dr. Skelton outlines the historical study of early maps in the past and in the present. Owing to Dr. Skelton's tragic death resulting from a car accident in 1970, an event that shocked the scholarly world, the manuscript for the book was unfinished and lacked full bibliographical references, delaying its publication until 1972.

William P. Cumming gave the second series of lectures in spring 1970 on the British cartography of eighteenth-century North America. Professor Cumming, already well-known for his wide-ranging knowledge of the historical geography and cartography of the eastern seaboard of the United States and Canada, reported and commented on the scholarly value of several new sources of manuscript maps relating to his theme. In the published version (1974), he covers the following themes: the cartography of the southern British colonies, the cartography of the northern British colonies, military cartography in the French and Indian War and the War of Independence, and the charting of the coasts, demonstrating the dramatic improvement from the *English Pilot* to the *Atlantic Neptune*.

With the 1972 lectures, the Director experimented with a different concept: that of starting with the subject matter that needed to be covered, and then selecting scholars who could provide a sound treatment of the needs and opportunities in that area. The topic chosen for 1972 was the history of map printing, in which six contributors distilled their basic research of the main historical techniques of map printing and presented their papers at a three-day symposium. The lectures were published in 1975 as *Five Centuries of Map Printing*, including such topics as the evolution of the relationship between cartographers and printers, the woodcut, copperplate and lithographic techniques, miscellaneous processes in the nineteenth century, and the application of photography to map printing.

The concept of a symposium was preserved for the 1974 lectures, in which Brian Harley, Lawrence W. Towner, and Barbara Petchenik presented lectures respectively on the eighteenth-, nineteenth-, and twentieth-century mapping of the American Revolutionary War, providing a comparative treatment of the contemporary and historiographic mapping of the War. The emphasis in both the lectures and the published version of the lectures was on the contemporary mapping of the war:

Dr. Harley's contribution on the eighteenth-century in the recently published book, *Mapping the American Revolutionary War* consumes three of the five chapters.

For the 1977 lectures, held immediately following the VIIth International Conference on the History of Cartography, the symposium concept was again retained, but the number of contributors was expanded to seven. The program, entitled "Mapping the Great Lakes Region: Motive and Method", included lectures delivered by an international group of scholars on the surveying and mapping of a varied physical region that has been affected by diverse historical influences. The papers ranged in subject from the pioneering maps of Champlain to the British, Canadian, and American efforts in hydrographic surveying in the nineteenth century. Plans for the publication of these lectures are now in progress.

A tentative program for the 1980 lectures has been planned around the theme "Art and Cartography," with the aim of combining the contributions of cartographers and art historians to this important but hitherto neglected subject. The lectures will follow the three-day symposium format and be accompanied by an extensive exhibition drawn from the collections of several cooperating institutions.

The symposium format of these lectures has succeeded in fulfilling its aims of providing publications of more general interest, but the constraints on the participants have tended to mould artificially the content and timing of their research, rather then encourage scholars to work on projects of their own design and for which they have set their own timetable.

In order to encourage independent research of this nature, the Nebenzahl Prize was established in 1977. This prize of $ 1.500 is for the best book-length manuscript on a topic within the field of the history of cartography. While it is not expected that a sufficiently large number of manuscripts will be forthcoming to award the prize every year, the Nebenzahl Prize Committee will make an annual evaluation. The prize includes automatic publication by the University of Chicago Press, with whose cooperation the award has been established.

Miscellaneous publications

The third series of publications sponsored by the Center, the Center's own series of occasional publications, also encourages independent research. Two books have already been published in this series: Kenneth Nebenzahl's *Bibliography of Printed Battle Plans of the American Revolution 1775 - 1795* (1975), and David Woodward's *The All American Map. Wax-Engraving and its Influence on Cartography* (1977).

Mapline, the quarterly newsletter, was started in March 1976 to provide short and timely information in the field of the history of cartography. It was decided at the beginning to make it strictly a newsletter; hence, its length has not exceeded eight pages and is unlikely to do so under present policy. Books reviews and articles are not included, although several short feature articles have regularly appeared. The publication attempts above all to be newsworthy by adhering to a strict quarterly schedule. The newsletter, edited by Patricia A. Moore, has achieved an international reputation as a valuable publication for libraries, professional cartographers, map collectors, map dealers, and all interested in this growing field.

Research and cataloguing projects

The continued growth of the Center depends upon its ability to attract outside support for projects in the history of cartography and related fields. These projects have so far been of two types: basic programs of cooperative cataloguing for printed maps of the Middle West, and basic research tools researching and displaying the historical positions of county boundaries. These projects will be discussed in turn.

Catalogue of Maps of the Middle West printed before 1900

The Midwest Map Catalogue project grew out of small feasibility study financed by the Council on Library Resources in 1971, when David Woodward visited some twenty-three collections in the Middle West to gauge the feasibility of a union catalogue of pre-1900 printed maps in Middle Western collections.

One of the main conclusions arising from this feasibility study was that there are two very different ways of compiling a regional cooperative catalogue. The conventional manner is for each participating institution to catalogue its own holdings pertaining to the region, and to report its holdings to the other institutions. This has two major drawbacks: (a) it does not take into account the maps held in depositories *outside* the area (such as the Library of Congress) and (b) it results in much duplication of effort; a common map might be catalogued by every institution.

The other method, which was adopted, is to assign part of the region to the institution that specializes in that region. In this case, the historical societies of each state in the region were charged with cataloguing the maps of their state. This is perhaps the first time that this principle has been used in the preparation of a cooperative map catalogue. The project has trained seven map cataloguers in a consistent system, given rise to much interstate cooperation in map cataloguing, and provided the cataloguers with the experience of working in other repositories than their own.

Between 1 February 1975 and 30 March 1977, cataloguers at the six cooperating state agencies (Ohio Historical Society, Indiana Historical Society, Michigan State Archives, Illinois Historical Library, State Historical Society of Wisconsin, and State Historical Society of Iowa) comp. catalogues of maps of their states. The Chicago list was comp. by the Chicago Historical Society, while the Newberry Library was responsible for listing maps of the Midwest and the Great Lakes Region as a whole.
In April 1977, the National Endowment for the Humanities made an additional grant to extend the catalogue to The West North Central States: Minnesota, North and South Dakota, Nebraska, Kansas and Missouri. When both projects are completed, the catalogue will cover the area known by The Library of Congress as the North Central States.

When these two catalogues are published by G.K. Hall in 1979 or 1980, the primary arrangement of the descriptions will be by area, subdivided by subject and date.
For example:

FOX RIVER VALLEY, WI - 1839
MENOMINEE RIVER, MI/WI - 1841
MILWAUKEE - VIEWS - 1881
WALWORTH CO., WI - LAND OWNERSHIP - 1873
WISCONSIN - 1857
WISCONSIN - GEOLOGY - 1847

Indexes at the back of the volume will list subjects shown on maps, surveyors, cartographers, draftsmen, printers and publishers named on the maps, and titles.

The aim is to make the catalogue as complete as possible. It will list separately published state, county, and city maps, including insets; state, county, and city atlases; Sanborn and other fire insurance maps of cities and towns; navigation charts of the Great Lakes Region from the early 17th century on; bird's-eye-views of towns; and maps in guide books, directories, histories, legislative manuals, state publications, the U.S. Congressional Serial Set, general atlases and general books. In addition, pre-1900 U.S.G.S. quadrangles will be indicated by index maps.

Besides the maps in their own institutions and in the collections of the Newberry Library, the individual cataloguers were urged to seek out maps of their areas in other repositories. All cataloguers examined the collections of the Geography and Map Division at the Library of Congress. Additional institutions visited included the University of Northern Iowa, University of Iowa, University of Illinois, Milwaukee Public Library, Michigan State Library, Michigan State University, Detroit Public Library, Indiana State Library, Western Reserve Historical Society, and the Cincinnati Historical Society.

The Midwest Map Catalogue will constitute a reference work of great value to historical and urban geographers, local and regional historians, librarians, genealogists, social scientists, historians of printing, publishing, cartography and the graphic arts, and workers in many other disciplines for whom maps represent a potentially important, if hitherto underused source. It is at that point that a third project will be initiated to convert the catalogue into a true union list of pre-1900 printed maps of The Middle West. At that stage, institutions will be asked to compare their holdings against the published catalogue, and to report their copies, paving the way for exchange programs to attempt to rationalize the resources of the region. It is anticipated that at that stage the technology and administrative organization of such comparative on-line data banks such as the OCLC (Ohio College Library Center) will have advanced to accomodate such a project.

The United States Historical County Boundary Data File Project

This project, funded by the National Endowment for the Humanities in December 1975, grew out of the recognition of the need for accurate county boundary base maps for historical research while the *Atlas of Early American History* was in preparation. In order to provide the maximum flexibility, it was suggested that the format of a data file be employed.

The goal was to create a machine-readable file of all the boundary changes of countries and other governmental and administrative units for the period 1790-1970. Used with a computer-directed plotter, the data file makes it possible to produce on demand maps of nearly any size, scale, and projection that cover any specified area (including interstate regions) at any given data from 1790 through 1970. The feasibility project covered New Jersey, Delaware, Maryland and Pennsylvania and was designed to demonstrate the concepts, methods, and utility of compiling such a boundary data file for the entire nation. In addition to county, state, and international boundaries, the demonstration project included all congressional districts, samples of state electoral districts and minor civil divisions, the locations of capitals and county seats, the areas of the spatial units, pertinent names, and the shoreline and major rivers of the two selected states.

The historical research followed the traditional approach of searching statutes, official maps and surveyors' reports, court reports, and other primary and secondary sources in order to establish a descriptive chronology of boundary changes from past to present. The research and draft cartographical compilations were executed at the Newberry Library under Project Supervisor John H. Long, formerly Assistant editor of the *Atlas of Early American History*.

Using the United States Geological Survey 1 : 250.000 topographic series as a base, all boundary changes were compiled on one map, and line segments were appropriately tagged to identify the time period they had been in effect. The lines were then digitized (converted into x,y coordinates) and a data file was designed and constructed by the staff of the University of Wisconsin Cartographic Laboratory under the direction of Joel L. Morrison.

An extension of this project will comprise a machine-readable data file of all changes in international, territorial, state, county and congressional district boundaries in the states of Old Northwest (Illinois, Indiana, Michigan, Ohio, and Wisconsin), 1788 to 1970. This project began on 1 October 1977 and will last two years.

Grants-in-Aid, and Research Associateships

In order to encourage the pursuit of independent research, the Center has offered five grants-in aid to scholars lasting from two weeks to three months (the maximum). Two research associateships have been awarded, one to Joel L. Morrison resulting in his *Recommendations for Classification of Extant Maps of the Great Lakes*, a report submitted in June 1975. In 1975, Helen Tanner worked for a month at the Center preparing her *Observations on Evaluation the Accuracy of Early Maps of the Great Lakes*, which she submitted in February 1976. These reports were of help in planning the 1977 Nebenzahl lectures, and may in addition form the basis of a proposal for a research project on methods of measuring the accuracy of early maps. This, and other projects, publications, and teaching programs now in the process of planning, will continue to develop the Hermon Dunlap Smith Center for the History of Cartography as a unique type of research and educational institute.

PUBLICATIONS OF THE HERMON DUNLAP SMITH CENTER FOR THE HISTORY OF CARTOGRAPHY

BERKMAN, Brenda: *Index to Maps in the Graff Collection of Western Americana.* Ed. by Robert W. Karrow, Jr. - Chicago: The Newberry Library, 1972.

CUMMING, William P.: *The British Cartography of Colonial America.* - Chicago: University of Chicago Press, 1974.

DUNCAN, T. Bentley: *Atlantic Islands, Madeira, the Azores, and the Cape Verdes in Seventeenth-Century Commerce and Navigation.* - Chicago: University of Chicago Press, 1972.

HARLEY, J.B.; Barbara Bartz PETCHENIK and Lawrence W. TOWNER: *Mapping the American Revolutionary War.* - Chicago: University of Chicago Press, 1978.

KARROW, Robert W., Jr.: *Mapping the American Revolutionary War.* Catalogue of an exhibition held at The Newberry Library. - Chicago: The Newberry Library, 1974.

KARROW, Robert W., Jr.: *Mapping the Great Lakes Region: Motive and Method.* An exhibition at The Newberry Library on the occasion of the Fifth Kenneth Nebenzahl, Jr. Lectures in the History of Cartography. - Chicago: The Newberry Library, 1977.

LAMB, Ursula [ed.]: *A Navigator's Universe. The "Libro de cosmographía" of 1538 by Pedro de Medina.* - Chicago: University of Chicago Press, 1972.

Mapline. A quarterly newsletter published by the Hermon Dunlap Smith Center for the History of Cartography at The Newberry Library. Ed. by Patricia A. Moore. No. 1, March 1976 to date.

NEBENZAHL, Kenneth: *A Bibliography of Printed Battle Plans of the American Revolution, 1775 to 1795.* - Chicago: University of Chicago Press, 1975.

QUINN, David B. [ed.]: *The Last Voyage of Thomas Cadendish, 1591-1592.* - Chicago: University of Chicago Press, 1975.

SKELTON, Raleigh A.: *Maps. A Historical Survey of Their Study and Collecting.* - Chicago: University of Chicago Press, 1972. Illustrated paperbound edition, 1975.

VIGNERAS, Louis-André: *The Discovery of South America and the Andalusian Voyages.* - Chicago: University of Chicago Press, 1976.

WASHBURN, Wilcomb E. [ed.]: *Proceedings of the Vinland Map Conference.* - Chicago: University of Chicago Press, 1971.

WOODWARD, David: *The All-American Map. Wax-Engraving and its Influence on Cartography.* - Chicago: University of Chicago Press, 1977.

WOODWARD, David [ed.]: *Five Centuries of Map Printing.* - Chicago: University of Chicago Press, 1975. Paperbound edition, 1976.

WOODWARD, David [ed.]: *The Hermon Dunlap Smith Center for the History of Cartography. Dedication Proceedings.* - Chicago: The Newberry Library, 1972.

MAP COLLECTIONS AND MAP LIBRARIANSHIP IN CANADA
REVIEW AND PROSPECTS

By Joan Winearls

Introduction

Smaller countries particularly those in the shadow of super-power neighbours tend to spend a great deal of their time examining the meaning of their nationality and its various support structures. Thus in Canada, we have recently been greatly concerned about the teaching of Canadian studies particularly at the University level. A recent report[1)] on this subject discussed at some length the teaching of geography and such support functions as map collections.

The potential importance of such enquiries for encouraging funding of map libraries and archives, which normally subsist on relatively modest budgets, must not be underestimated. The sections of the Symons Report which bear on this topic highlight key aspects of map libraries as they should be:

"A great number of good, comprehensive map collections should be established at universities and colleges and existing collections should be strengthened. There should be several major depositories for current maps and one for historical maps at universities in each of the regions of Canada. Such collections should develop regional and thematic specialties, ... Universities can provide a valuable service to their respective communities, ... by serving as centres for cartographic information about their own particular regions, ...

Information about their map collections and about the potential usefulness of these collections should be made more widely known by universities to their students and teaching staff, and to other educational institutions in the community The scope of maps as resource materials for Canadian studies at all levels of education and in a wide variety of disciplines should be emphasized. Instruction in the use of maps should be included in teacher training programmes and systematic collections of maps should be developed in every high school as a normal element of the school library. ...

Good map collections are, if anything, more difficult to develop and maintain than good collections of books. Map collections should, therefore, be made the responsibility of trained personnel who are familiar both with library practice and with the academic disciplines served by such collections. To this end the place of map librarianship should be further recognized by schools of library science, and the pioneering work of the University of Toronto in training map librarians should be taken as an example for the development of similar courses elsewhere. ...

More active programmes both to collect and to catalogue maps are required. In particular, there is a need for a comprehensive, national catalogue of maps that will provide to all researchers in the social and physical sciences, in one place, a knowledge of what is available throughout Canada. Fortunately, the projected National Union Catalogue of Maps, which is being prepared under the joint direction of the Association of Canadian Map Libraries and the Public Archives of Canada, will meet this requirement. However, progress in the creation of the National Union Catalogue of Maps has been slow, primarily because of lack of money. The Catalogue has, in fact, been under discussion since 1967 and under preparation since 1973. The Commission urges the Canadian Government to provide the funds needed to complete this important project at an early date. ...[2)]

The extent to which parts of this ideal have been approached or been realized will be examined in this general review of the state of Canadian map collections and map librarianship.

Map Collections

There is very little written about Canadian map collections either as a group or individually and certainly many collections have been in existence long enough that useful articles could be written on the trends in their growth and development. As well an overall analysis of collections would be useful in pointing out directions for development in the next two decades. Some information for comparison, however, can be derived from the first and third editions of the *Directory of Canadian Map Collections.*[3] The factors that can be compared from these directories are as follows: number and type of collections, size, rate of growth, number of staff, hours; and to a lesser extent acquisition policies, cataloguing and reference services. Date and circumstances of origin, physical facilities and equipment can be derived from other sources.

The date of origin of many map collections in Canada is difficult to establish partly because the collections themselves often report different dates of establishment according to varying concepts of when a collection begins: with full-time staff?, with its own room?, with a proper budget and collection policy? For instance, of the twenty-one Canadian collections listed in the *World Directory of Map Collections,*[4] eleven are listed as being established prior to 1960, while an article by Sauer[5] clearly shows all libraries being established as *separate entities* in the 1960's. Major archives such as the Public Archives of Canada and public libraries were the earliest collecting maps although in some collections the items were part of other gifts or transferred from government agencies and often little was done with the maps except to provide storage. The university geography departments in the pre-1960's era often had collections of ten to 30 000 maps but these were working collections for field trips and class work and the maps were often "used up". In both cases, however, the real first or major growth period in Canada came in the 1960's when, as universities expanded, many geography departments established their map collections properly with a full-time professional as opposed to part-time graduate students in charge. At the same time the National Map Collection also increased sharply in size and staff in the 1960's from 139 000 sheets, and five staff in 1959 to 450 000 sheets, and twelve staff in 1969. A move to a new building with more space available and an increase in the restoration programme also accompanied this growth. Unfortunately, other archives and public libraries did not see the same need for increase in interest, size or care for the map collection, at that time; recently, however, several provincial archives appear to be developing their map collections more extensively. For instance, in comparing the institutions in the two directories of Canadian map collections, eight archives were newly listed in the 1977 directory compared with twelve listed in both directories. On the other hand the strong growth is still with the university map collections as twenty-five new university map collections were included in the 1977 directory compared to twenty-eight in both directories.

There are still few significant collections in public libraries, only three are compared between first and later editions of the directories and two more are newly listed. There are also few true map collections in government agencies: six government collections being compared in the first and third edition of the directories while seven were added. No map collections are listed for businesses, or school libraries. Admittedly any collections in those institutions would normally be small working collections but it has been observed that school libraries in Toronto for instance seldom include maps.

In toto forty-nine collections can be compared in the 1969 and 1977 directories, forty-three map collections were newly reported and thirty-one were deleted. Of those deleted some particularly in the universities were probably amalgamated with other collections.

Except for the National Map Collection, the larger collections in Canada have grown steadily since 1969 but are still at or under 200 000 maps.[6] Considering that university libraries in Canada did not begin their collections with enormous numbers of United States Army Map Service maps as did many of their American counter parts, this seems to be a fairly healthy situation. The larger collections over 100 000 (University of British Columbia, University of Alberta, University of Western Ontario, Canada Dept. of Energy, Mines and Resources, Université d'Ottawa, University of Toronto, McGill University) appear to have achieved this with varying rates of growth from 3 750 to 12 000 maps per year. Three university libraries have grown very quickly (University of Alberta from 50 000 to 200 000 maps; Université d'Ottawa from several hundred to 102 000 maps; University of Western Ontario from 31 000 to 129 000 maps) which suggests either much higher budgets than for the other libraries, or large gifts or deposits of maps. Budget information is, however, generally not available for most libraries. Most libraries show a fairly steady overall growth rate in number of maps more or less proportional to the size of the collection. S. Sauer notes that in Ontario "the trend is for a new library to develop at an accelerated pace for the first three to five years and then to reduce acquisitions to approximately 50-60 percent of the initial yearly expansion".[7] While this may have been the situation with Western and Ottawa, the figures in the two directories for over-all size and accessions appear to show a more even growth rate. The three large collections mentioned earlier all show larger acquisition rates than the other collections of 10 000 to 12 000 maps per year, and most of the other libraries show an increase in acquisition rate as their collection has grown.

The hours of opening for libraries have not varied greatly in the decade. Generally, university collections have the longest hours, although almost all have cut back slightly in the period. Average university hours are around 60-65 hours per week while government and archival collections tend to be open about 40 hours per week. It is interesting to note that the three largest institutions: The National Map Collection, Canada Dept. of Energy, Mines & Resources map collection and University of Alberta map collection are open for the shorter group of hours, a situation which does not seem to correlate with the purpose in establishing such large and important collections.

In about fifty percent of the cases staffing for all collections has either stayed the same or increased very slightly. This is unfortunate considering that all collections have grown and certainly require more time to be spent on care, cataloguing, processing and especially on reference work. Except for the National Map Collection where staff increased from thirteen to twenty-four only one large collection has a staff of more than 4,5 including student assistants (Université d'Ottawa with 7,5). Most other collections vary from one person to two full-time staff at the most.

There is a considerable variety in number of depository arrangements for government maps both Canadian and foreign from library to library. For those reporting on this (mostly university collections) the number varies from five to twenty-eight. Many of the large libraries, however, are depositories for fifteen agencies. Certainly for small map libraries in Canada, the emphasis in acquisitions is often still on what can be acquired free. The National Map Collection's exchange arrangement in providing superseded foreign topographic maps is very important in providing not only free maps but series that are not available elsewhere. Rising costs of foreign maps and the fact that more Canadian government agencies particularly at the municipal level are changing for maps has added considerable pressure to acquisitions budgets. Specializations of university map collections appeared to have changed somewhat between the two directories but it is doubtful that this is often the case and may depend on different wording of the questionnaire in the two directories.

Cataloguing systems in libraries have changed little in the period between directories. Map collections appear to be using the same variety of in-house and other systems now as they were before. Various levels of completeness in cataloguing were recorded. This situation probably reveals the desire to wait for the national map cataloguing code and system before making changes or going further into automation.

Many map collections moved to larger quarters in the decade of the 1970's and a very good sample of the spatial layouts and facilities is obtained from a publication showing plans of eight mainly larger university map libraries in Canada.[8)] The space varies from 2 100 square feet for a collection of 65 000 maps to 10 152 feet for a collection of 130 000 maps but there is little correlation with size, for example, the University of Western Ontario with a collection of 129 000 maps has only 4 900 feet, scarcely sufficient for expansion. The layouts are completely varied as most map libraries must fit into various networks of permanent rooms, corridors etc. Generally, one-quarter to one-third of the area is reserved for users, at least one-half for storage and the rest is working area for staff. The majority of the storage area is open access in all libraries. Most university map libraries provide a reasonable number of light tables, photogrammetric equipment and an enlarger-reducer and a photocopier.

The archival collections have lagged behind in acquiring equipment usually because of legitimate fears of misuse of valuable maps.

Little is known about reference service in the various libraries. The directories did indicate an increase in publications among map collections and particularly of orientation guides, special map lists and atlas lists as opposed to the earlier vogue for accessions lists. Most libraries except archives loan their maps, most have changing displays and some will have orientation programmes. But further study should be done on the types and numbers of users, types of questions, maps most frequently used and the nature of reference work in all types of collections.

National Map Collection

The growth and activities of the National Map Collection has been documented in the annual reports of the Public Archives of Canada and the annual reports presented to Association of Canadian Map Libraries. Though collecting maps from 1872 the important growth points came in 1908 with the establishment of a separate map room based on the transfer of the large military 'C' and Ordnance series, again in 1949 with the decision to collect current Canadian maps as the best way to establish a good archival collection and again in 1968 when the collection became known as the National Map Collection.[9)]

In 1965 150 000 foreign topographic maps were transferred from the Department of Energy, Mines & Resources serving as the basis for establishment of a foreign section. The National Map Collection now has a collection of 750 000 maps with current accessions of about 50 000 sheets. Most maps are received as current production or transfers from the Canadian government, on deposit from provincial governments, by request and purchase from other sources or on exchange from foreign governments. In the latest directory the collection lists the largest number of depository arrangements of any library in Canada (twenty-eight) but only fifteen percent of the collection is foreign. The National Map Collection attempts to acquire all Canadian maps published and prepares Canada's contribution for *Bibliographie Cartographique Internationale*. Recently the card catalogue of 97 800 cards of the Canadian Section was published[10] making information on the country's largest collection of early and recent Canadian maps available to various libraries across Canada and elsewhere. The section has also recently published over a dozen bibliographies of maps and atlases (e.g. *County maps, Winnipeg in maps, French atlases in the rare atlas collection*) as part of the general diffusion of archives policy of the Public Archives of Canada. Recent exhibits on birdseye views and mapping of the arctic are also part of this programme; the latter was produced in copy form so that it can travel.

The National Map Collection leads the country in providing information on current preservation practices for maps and in developing other programmes for long-term preservation and dissemination of information through 105 mm microfilming of the entire collection. The latter was begun in the summer of 1977 and is expected to take several years to complete.

The National Map Collection may be unique in the world of national collections in that it combines the national archival function and national library function for maps. This creates certain difficulties as the institution resides within the Public Archives of Canada. Staff is thus almost totally from the archivist background and until recently with academic background in history rather than geography. As well no positions for librarians have yet been established. For the Canadian requirements the collection operated very well combining the typical library function of acquiring, processing and cataloguing current and early printed Canadian maps with the archival function of treating transfers of government map records by provenance. A re-distribution programme for sending older duplicate Canadian maps to Canadian map libraries is now being planned and should be operational soon. The operation of a national foreign collection, however, is more of a library function and here the National Map Collection is inhibited by operating under the Public Archives of Canada Act, which does not really provide for the acquisition of foreign materials. Thus the growth of the foreign collection is very slow and has basically not proceeded beyond the exchange of topographic sets with foreign countries. It can be argued that a reasonable national collection of foreign maps is needed in any county and this would help to co-ordinate the extensive foreign resources in university map collections. The dispersal of superseded topographic sheets in the section's redistribution programme to other Canadian map libraries is very useful to these libraries but is criticized by map librarians since it does not ensure a national foreign archival collection for historic research in the future.

Association of Canadian Map Libraries

There is little doubt that the single most significant development in map librarianship in Canada was the establishment of the Association of Canadian Map Libraries in 1967. In a large country with a relatively small population there are still under 100 map collections and even many of the major collections have still only one professional on staff. There is, and was, an overwhelming need for these people to be brought together to work on common national problems in map libraries and archives. But the fact that the Association is stronger to-day than ever before and has a reasonable number of accomplishments to its credit shows what a small dedicated group can do.

At the inaugural meeting arranged by four map librarians in Ottawa in June 1967, T.E. Layng, head of the National Map Collection, set the tone for the next decade's work by suggesting that Canada as a young country was in a good position since no map collection had grown beyong the point of no return in terms of descarding older ineffective methods to reach common ground on new, particularly vis à vis a national map cataloguing system.[11)]

The association established several patterns early in its existence which have continued since. Active membership in the association has been limited to map librarians working in Canada, though there is also a category for associate members from any country or profession. This restriction, though somewhat narrowly nationalistic in outlook and much discussed over the years, has been to ensure that we look after the main goal first: the development of Canadian map libraries and map librarianship. The group has been fairly evenly divided between librarians, geographers and archivists. Though some clashes have resulted from this mix, generally it has served to keep members open-minded and flexible in their attitudes to various projects. A tradition of annual meetings alternating generally between east and west was quickly established and attendance has been between thirty and seventy at each meetings. As with other map library associations papers on all aspects of map librarianship, map production, use and analysis have been presented at these meetings. R.I. Ruggles noted that "the annual conferences have provided one of the few forums in Canada for the presentation of papers dealing with early maps".12)

A review of some of the projects of the Association over the years will highlight some of the accomplishments and concerns of Canadian map librarians.

The Association has been reasonably active as a publisher. For the first ten years the main journal publication was the edited *Proceedings* from each annual conference. A *Newsletter* was begun in 1968 slowly growing in size and changing its name to the *Bulletin* with issue no. 16 in 1974. In 1977 it was decided that the *Bulletin* which had more varied contents such as reviews, news notes, map lists as well as articles and could be issued three or four times a year, would become the major publication and that the *Proceedings* would be discontinued. The association has also published two directories of map collections as noted before which have been basic documents for the location of map resources in the country. In 1975 the Association published a folio of plans of new map libraries as a contribution to the problem of physical layout of map collections.13) The association has also ventured into the field of facsimile map publishing in an effort to make important early maps more widely available. This has been done with the assistance of sponsors to help underwrite the cost and five maps have appeared to date.

Standards and procedures for map libraries and the question of education for map librarians have been of great concern to members from the beginning. The idea of publishing a manual was discussed as early as 1968 and papers on planning, equipment and exhibits were actually published for comment in the Association's *Bulletin*. After some years of discussion about the nature of such a manual the association has finally decided to prepare a manual for small map libraries which should be published in the near future.

In terms of education generally the association has provided informal training for new map librarians by encouraging them to take part in the activities of the association, attending conferences and workshops and writing articles and reviews for the *Bulletin*. But the Association has not moved to set up courses in education for map librarians partly because it could not give academic credit for such courses. Moreover it can be argued that the role of education for map librarianship in Canada should be in the hands of the professional training schools for librarians and archivists which have the resources to finance the programme and particularly when they can use practicing map librarians and archivists to conduct the courses. To date only the University of Toronto Faculty of Library Science operates a course on Map Librarianship. This course has been given five times since 1970 and over 100 students have taken it. Though there are few full-time positions for map librarians in Canada many of the students took the course because they were working or were about to take positions in public, business, school or government libraries where they often have had to cope with small map collections. The Public Archives of Canada has provided training in archival map management in the past in its various training courses for archivists. In the spring of 1978 the National Map Collection is also planning a conference for provincial and other map archivists which will provide some training but also discuss the co-ordination of map acquisition policies among archives. Perhaps this will lead to more regular training sessions for map archivists.

The association has made considerable efforts to affect national policies that involve maps and have met with some successes and some failures. The association for instance has very good relations with the national map-producing agency the Department of Energy, Mines and Resources and from 1971 on has heard annual reports from the department at the conferences, with opportunities for the members to question and comment on policy. Recently when the educational discount for topographic maps was discontinued the Association was able to lobby effectively to have part of this discount renewed.

One of the great areas of concern has been the need for a national current bibliography of Canadian maps which must be based on comprehensive national collecting through the enforcing of the deposit law for maps. At present the National Map Collection which provides the Canadian contribution to *Bibliographie Cartographique Internationale* estimates that it only catches 50 - 75 percent of annual Canadian map production. In 1970 the Association requested that the Secretary of State for Canada enforce the deposit law for maps and transfer the control of this from the National Library to the National Map Collection. Unfortunately, the government is still considering the intricacies of this problem and the law is not being enforced. Thus as with many other countries Canada is still far from being able to produce this important tool.

The Association has established various committees from time to time to deal with common problems. Two of the most important at present are the Conservation committee and the Microfilming committee. The Conservation committee has done excellent work in organizing joint orders for acid-free map folders, and providing information and testing on new products such as tapes. The committee has also surveyed current conservation practices in map collections as a basis for standards and is compiling a bibliography of articles in the field The microfilm committee is maintaining liaison with the National Map Collection in their current 105 mm microfilming project and gathering information on readers, reader-printers and costs of filming other groups of maps. With rising costs for space and the evident suitability of large 105 mm format for certain large scale series and for the conservation of early maps, there is a growing interest in the possibilities of this format among Canadian map librarians.

From time to time groups of members have held regional meetings between annual conferences but the establishment of chapters has never been formalized within the Association. In some areas local co-operative systems may take over this function. The Ontario University Library Cooperative System was set up in 1973 mainly to operate co-operative cataloguing projects to save money and avoid duplication of effort in Ontario university libraries. After some discussion the map group established its main project as that of assistance in the development of and contribution to the National Union Catalogue of maps. Since this is not yet operational the group has gone ahead with other useful projects: such as the publication of a union list of atlases in Ontario[14)] (following the lead of some British Columbia and Quebec map libraries), the establishment of procedures for inter-library loan of maps, a map exchange system and the initiation of the revision of a union list of topographic series in Canada first published by the National Map Collection in 1973-4.[15)] Other local meetings to solve common problems have been held in Quebec and Vancouver and are perhaps more useful than chapter meetings in that they are project-oriented.

The Association of Canadian Map Libraries became a member of IFLA soon after the Geography and Map Subsection was organized and sends a delegate regularly to meetings. Members have also been involved in the directory committee and the education committee. The Association is also maintaining liaison with the Special Libraries Association Geography and Map Division and the Western Association of Map Libraries. Though little concrete has yet emerged, several joint conferences have been held, members from each group attend each others conferences and a member from Association of Canadian Map Libraries is on the Special Libraries Association Geography and Map Division's Committee on Standards. Nevertheless the difficulties in effective liaison with such a large group of people spread over a large continent have served to reaffirm for the Association at the ten year anniversary the need to maintain its autonomy.[16)] As well the maturity of the Association seems evident in the active role many members play in the new Canadian Cartographic Association and the fact that Association of Canadian Map Libraries is one of the five association members on the National Commission for Cartography one of whose important roles is to represent the Canadian cartographic community in the International Cartographic Association.

National Map Cataloguing[17)]

At the first meeting of the newly-forming Association of Canadian Map Libraries in 1967 one of the main objectives of the group was to investigate the possibility of a national union catalogue of map holdings. Basically, this was an attempt to solve both the national documentation problem and the cataloguing problem for maps as quickly as possible. The association established a National Union Catalogue of Maps Committee which was to look into all aspects of the problem including use of the computer, amount and type of cataloguing information required, classification systems, area names and subject terms. In the first few years various data was collected, systems were investigated and the development of MARC at Library of Congress was followed with interest. In 1969 the National Union Catalogue Committee, in an effort to get some results, designed a National Union Catalogue Reporting Programme. The programme was entirely manual and was based on a 5 x 8 card which had positions for 22 pieces of information. Brief rules for cataloguing (never intended to be comprehensive) were also established. Libraries across Canada purchased cards, filled them in with reports of their acquisitions and sent them in to Ottawa. The system was intended to test response and it was presumed that the data could be converted later into a machine system. The latter point proved, of course, to be illusory.

Initially the response to the Reporting Programme was quite good though it dropped off as it became obvious that there were inherent problems in the system and the main thrust of development moved towards machine-readable systems. In a sense the value of this "pilot project" was in educating Canadian map librarians (many of whom have had no training in cataloguing) to the problems of map-cataloguing. But it soon became apparent that the brief rules were far <u>too</u> brief and the cards sent in were inconsistent, many different entries often being received for the same map. The need for detailed rules thus became evident. At the same time in comprehending the amount of work involved in map cataloguing, most of it after all being original cataloguing, many map librarians felt the need for emphasis also on <u>shared</u> cataloguing within any National Union Catalogue programme.

By 1971 it was realized that no group of volunteers from a small association would have the time, money or expertise to develop the National Union Catalogue System at a rate that would be fast enough. Accordingly, the Map Division, Public Archives of Canada was approached to take over the work. Their reply was cautious at first because as an archive they were only beginning to investigate the area of automation. The 1970-71 chairman Yves Tessier then recommended that the Association ask for financial aid and expertize from the National Library, Research and Planning Branch, the section that was responsible for the development of bibliographic projects and co-operative projects between libraries in the country. The Association did not agree to do this but the emphasis in the report for that year on the need for assistance from an appropriate institution had the effect of encouraging the Public Archives of Canada to take stronger action. In late 1971 they agreed to fund meetings of the National Union Catalogue committee and were working on the establishment of a new position for a map cataloguer to develop an automated national cataloguing system for maps.

In 1971 the long awaited *MARC Map Manual* was published by the Library of Congress. This was a crucial development for the National Union Catalogue as it was likely that there would be national support for any MARC system, and it gave the committee something concrete to consider. The Manual was studied in detail at the 1972 working meeting of the National Union Catalogue Committee with Hugo Stibbe as chairman and the members came to the conclusion that the structure of the format was eminently suitable for Canadian application. However, there was some disappointment about the lack of certain elements in the format such as provision for an area main entry, co-ordinates, and a coded scale among others and the Committee decided to use it only as the basis for the development of a *Canadian* MARC-map format.

By mid-1973 the Public Archives of Canada had hired H. Stibbe to develop automated cataloguing procedures in the National Map Collection, and to work on the National Union Catalogue of Maps project, enabling the project to proceed more quickly Though the developments of the map formats were now firmly entrenched in the Public Archives of Canada, there has been growing co-operation with the National Library stemming from their extensive involvement in MARC formats, and cataloguing standards generally. To date there are agreements in several areas: Map records will be created in the National Map Collection and National Union Catalogue but the processing of these records will be done by programmes and systems developed by the National Library. Author headings for maps will be established by the National Library, while area name entries established for maps by the National Map Collection and National Union Catalogue could be used for subject entries in the National Library for monographs and serials where applicable. The National Library has created a common authority file format in which such elements as geographical names, topical subject headings, personal and corporate names and others, in both the English and French language, where required, are established. The decision for co-operation in establishing such headings will eventually affect the Canadian MARC Communications formats.

In 1972 and 1973 H. Stibbe designed a first draft of the Canadian MARC format for maps. This was reviewed by the National Union Catalogue Committee in February 1973. The main changes in the format include the elements which were not present in the MARC II Format for Maps of Library of Congress as mentioned earlier. In April 1975 this draft Canadian MARC for maps was formally turned over to the Canadian MARC Office in the National Library who will produce the authorized version compatible with other Canadian MARC formats.

Because of the inadequacy of the Anglo-American Cataloguing Rules for cataloguing maps as well as other map codes now badly out of date the National Union Catalogue Committee decided to create a complete set of descriptive rules for maps alone to be based on ISBD principles and to construct rules for main entry by area. The Canadian Cataloguing Rules for Maps (descriptive section) were drafted by H. Stibbe during 1973 and 1974 and were considered in detail by the National Union Catalogue in April 1975. Before this meeting two libraries had been using them to catalogue several hundred maps, the University of Ottawa and the University of Toronto Map Library. This produced information on some problem areas and together with the advice of other practising map cataloguers some revisions were made. The rules were then put aside until the ISBD (Cartographic Materials) was also in draft form in case there was a conflict.

The International Standard Bibliographic Description (Cartographic Materials) was developed during two meetings of the Joint Working Group ISBD (Cartographic Material) set up by the IFLA International Office for Universal Bibliographic Control - one in 1974 and a second in September 1975. The Chairman, H. Stibbe, then prepared a draft which has been circulated to Committee members, other ISBD Committees and other map librarians and has gone through several revisions by the committee. The ISBD (CM) covers the description of maps, globes, atlases, aerial photos, etc. and prescribes both compulsory and optional descriptive elements a citation order, and the demarcation of elements by a punctuation code. Entry is not part of the format. The prescribed elements and layout naturally affect national cataloguing codes though the ISBD's only define generally how the element is recorded in each case. Rules for the construction of these elements are thus the concern of national cataloguing codes.

At the annual National Union Catalogue meeting in April 1976, the Draft Canadian Cataloguing Rules for maps were revised to include ISBD provisions and punctuation. On May 1st the Draft Rules were submitted to the Revision Committee of the A.A.C.R. for consideration for the second edition through the Canadian Cataloguing Committee, one of the Joint authors of the A.A.C.R. In March 1977, the National Union Catalogue Committee was asked to examine the draft chapter for maps in the second edition A.A.C.R. and sent in a joint statement with the Library of Congress Geography and Map Division and the Ohio College Library Center automated map cataloguing project to the A.A.C.R. revision committee. Though it can be said that the Association of Canadian Map Libraries has had some influence on the revision of the A.A.C.R. chapter for maps, it will not be known until publication whether this code can be easily used for map cataloguing.

In the ten years development of the National Union Catalogue there has admittedly been no creation of machine-readable records. But considerable advances have been made: the creation of a very detailed cataloguing code which has the capacity for use with both early and modern maps, a major contribution in the creation of ISBD (CM) in the person of the chairman H. Stibbe and the support of a national institution in the creation of national map cataloguing. The project will go on being co-operative; despite government budgeting the National Map Collection cannot provide all the cataloguing alone, nor can many map collections provide the manpower to catalogue all their maps and certainly regional groups such as the Ontario University Library Cooperative System (Map Project) will play an important part in helping to develop the logistics of the co-operative aspects.

Prospects

The analysis of Canadian map collections and map librarianship in the last decade shows clearly that the speciality in Canada has come of age. Many collections are firmly established with several years development and experimentation behind them, many Canadian map librarians and curators between them today have an average of ten years experience in the field, enough to have a clear idea of what is feasible for their map libraries. Thus though budgets are restricted these people are in an excellent position to continue to improve the quality of their services, where extensive manpower or money is not needed. Though a major cataloguing effort is still ahead of most of us, the fact that many collections have been established for over ten years means that many of the housekeeping tasks in terms of the maintenance of collections have been completed, or have been organized as part of an ongoing programme.

In commenting on the statements in the Symons report it should be noted that many of the ideals cannot be achieved if budgets are severely cut back. The information in the directories suggests that many new map collections have been established in universities though not interestingly enough in many of the community colleges. Most university libraries have a reasonable number of depository arrangements for federal and provincial government maps but perhaps the report will put pressure on municipal governments to consider deposit arrangements to local map collections as part of their public policy. Local collections of historical maps are already being developed in provincial and city archives and various rare materials collections but many of these archives and libraries are slow to provide staff and funds to speed up this process. Collections of photocopies or 105 mm microfilm copies of maps held in the National Map Collection could be deposited in a local University collection to complement other local holdings and perhaps the federal government could assist in this. University map collections already serve as community centres for cartographic information for both Canadian and foreign maps but on increasingly small university budgets. As well the large university map collections have a double burden in supporting the major collections in the country of foreign thematic maps and atlases as well as substantial collections of Canadian maps. The National Map Collection could assist by creating a larger national collection of foreign maps and other groups in the community such as business, local government and private donors could assist by donating to local collections. Moreover, if budgets continue to go down there may have to be more co-operative acquisition arrangements among map libraries and an increase in the use of inter-library loan services.

Generally, map collections are becoming better known in their local area as they grow in size, services and in age. Many university libraries are also doing more advertising and offering orientation programmes, and most are now used by many departments other than geography. Use will always depend on the one hand on effectively organized and staffed collections and on the other the amount of teaching of map use to students in schools and universities. "Map literacy" is still not common knowledge in Canada or anywhere else. One cannot but hope that the Symons report will encourage the several other library schools in Canada to develop map library courses. Most Faculties of Library Science have experienced map librarians locally available to teach it.

Finally the Symons report draws attention to the importance of the National Union Catalogue of Maps Project. Although considerable progress has been made, once again it is clear as it was in 1971 that developments cannot go on at such a slow pace. It is now time for the government to increase the manpower devoted to this project in the National Map Collection from two to a sufficient number to put it into the operational phase within a two year period.

NOTES

1) T.H.B. SYMONS: *To Know Ourselves: Report of the Commission on Canadian Studies. Ottawa 1975. Referred to hereafter as the Symons Report.*

2) *Symons Report, ibid vol. I, p.58, 60.*

3) J. WINEARLS; Y. TESSIER: *Directory of Canadian map collections, 1969;* L. DUBREUIL *(comp.); Directory of Canadian map collections, 3rd ed. 1977.*

4) W.W. RISTOW [*ed.*], *World Directory of Map Collections, 1976.*

5) S. SAUER: *University Map Collections in Ontario: new trends and development. In: SLA, Geography and Map Division. Bulletin, 90, (1972), p.20.*

6) *The National Map Collection because of its much larger size and uniqueness is generally not compared in this section but is discussed separately.*

7) S. SAUER: *University Map Collections in Ontario, ibid., p.21.*

8) S. SAUER: *University Map Libraries in Canada: a folio of selected plans, 1975.*

9) *National Map Collection. In: National Library Newsletter, 1, 1969. p.9-10.*

10) *National Map Collection, Ottawa. Catalogue of the National Map Collection Public Archives of Canada, Ottawa, Ontario. Boston: Hall, 1976. 16 vols.*

11) *Association of Canadian Map Libraries. Proceedings of the annual conference. 1, (1967) Ottawa, p.6.*

12) *Association of Canadian Map Libraries. Proceedings, ibid 10, (1976), p.15.*

13) S. SAUER: *University Map Libraries in Canada, op. cit.*

14) K. DONKIN; R. FINCH: *Union List of Atlases in Ontario Universities. Toronto: Council of Ontario Universities, 1976. 253 p. 3193 entries. The latest issue of the Quebec list came out as a companion volume:* Y. TESSIER: *Catalogue collectif des atlas des cartothèques universitaires du Québec. Quebec, 1976. 253 leaves. --* R. MACDONALD: *Union list of atlases, Provincial Archives, Simon Fraser University, University of British Columbia. University of Victoria, Vancouver City Archives, Vancouver Public Library.* [*Burnaby, B.C.*] *1974. 155 p. 1862 entries.*

15) *National Map Collection, Ottawa. List of map series in the Foreign Section. Comp. by Vivien Cartmell. Ottawa: National Map Collection, Public Archives of Canada, 1973-4. 4 vols. This union list which records holdings of 17 collections beside NMC has been very important in Canada in providing information for inter-library loan and in vividly demonstrating by the inconsistencies of its entries the need for proper international cataloguing of topographic series.*

16) *ACML Bulletin, 21, (1976), p.3.*

17) *Much of this section is drawn from H.L.P. Stibbe, MARC-Map: The history of its development and a current assessment, Utrecht 1976; from discussions with Dr. Stibbe and from annual reports of the NUC Committee.*

BIBLIOGRAPHIC REFERENCES

Association of Canadian Map Libraries. *Bulletin (formerly Newsletter)* Ottawa etc. , no. 1 (1968)

Proceedings of the annual conference. 1-10, 1967-1976. Ottawa, etc.

National Union Catalogue Committee. Reports, 1967-1977. (In: *ACML Proceedings, ACML Bulletin* and association files.)

Canada. Public Archives. *Reports*. Ottawa 1881-

DUBREUIL, Lorraine [comp.] : *Directory of Canadian map collections, Repertoire des collections de cartes Canadiennes*. 3rd ed. - Ottawa: Association of Canadian Map Libraries, Association des cartothèques canadiennes, 1977, 126 p.

KIDD, Betty: Report from the National Map Collection. In: ACML *Proceedings* 9th Conference, (Sackville 1975), p.83-89, and 10th Conference, (London, 1976), p.59-63.

National Map Collection. In: *National Library Newsletter*, vol. 1, (1969), 4. p.9-10.

Ontario University Library Co-operative System. Map Project. Minutes of meetings 1973- [unpublished] .

RISTOW, Walter W. [ed.]: *World Directory of Map Collections*. Compiled by the Geography and Map Libraries Sub-Section IFLA. - München: Verlag Dokumentation, 1976. (IFLA Publications. 8.)

SAUER, Serge A.: University Map Collections in Ontario: new trends and development. In: Special Libraries Association. *Geography and Map Division Bulletin*, 90, (1972), p.20-24.

SAUER, Serge A.: *University Map Libraries in Canada: a folio of selected plans*. - London, Ont.: Association of Canadian Map Libraries in cooperation with the Department of Geography, University of Western Ontario, 1975. 1 portfolio.

STIBBE, Hugo L.P.: *MARC-Maps: The history of its development and a current assessment.* - Utrecht and The Hague: Geographical Institute of the State University and Netherlands Organisation for Information Policy, 1976. 292 p.

SYMONS, T.H.B.: *To Know Ourselves: Report of the Commission on Canadian Studies.* - Ottawa: Association of Universities and Colleges of Canada, 1975. 2 vols in 1.

WEISSMILLER, D.: Do we need a national bibliography of Canadian maps? In: ACML *Proceedings of the annual conferences,* 6, (1971), Toronto, Ottawa, 1972, p.15-26.

WINEARLS, Joan; Yves TESSIER: *Directory of Canadian map collections. Répertoire des collections de cartes Canadiennes.* - [Toronto]: Association of Canadian Map Libraries; Association des Cartothèques Canadiennes, 1969. 72 p.

GEOGRAPHY AND MAP LIBRARIES IN DENMARK

By Ib Rønne Kejlbo

In the Middle Ages practically all centres of learning in Europe were attached to the church. It is therefore not surprising that the first Dane who influenced the opinion of the learned world regarding the geography of the Scandinavian countries, was himself an ecclesiastic. The maps of Claudius Clavus, dating from the first half of the 15th century, were made as a supplement to Ptolemy's works and may be characterized as an interpretation of the Northerners' voyages in the North Atlantic Sea, whether deliberate on Clavus' part or not. For a long time his maps of the Scandinavian countries and Greenland manifested themselves in the geographical views held by European scholars.

The next Dane to supply new cartographic information on Scandinavia, or rather Denmark, was Marcus Jordan, 1521-95, professor of mathematics at the University of Copenhagen. Through the Duke Henrik Rantzau Jordan came into contact with the publishers Georg Braun and Ortelius, and in this connection it is worth mentioning that the oldest extant printed map of Denmark is to be seen in Georg Braun's *Civitates Orbis Terrarum*, Vol. 4, 1587.

During the same period another Dane devoted his time to cartography, and that was Tycho Brahe, 1546-1601. One evidence of his studies is the map of the isle of Vén in his work *Epistolarum Astronomicarum Libri*, 1596. The map was based on triangulations. Unfortunately, Tycho Brahe did not complete the triangulation net of Denmark as he had intended to.

In the years from 1600 to about the middle of the 18th century various efforts were made to produce a national map in print, but without much success. Nothing epochmaking happened until 1742 when The Royal Danish Academy of Sciences and Letters was formed. Thanks to the Academy, the first mapping of Denmark on a scientific basis was then initiated. The surveying took place from 1757 to 1820.

Geography as taught subject in Denmark

The University of Copenhagen was founded in 1479, but about a hundred and fifty years passed before scientific disciplines came to be taught here, of which geography and botany were the first, and even then education in geography was given in conjunction with history. The first professor of geography was appointed in 1635, however, the first independent Chair of geography was not set up until 1888. Nowadays, the education of geographers is offered not only at the University of Aarhus, where the first professor of geography was appointed in 1943, and - since 1964 - at Danmarks Laererhøjskole, Copenhagen (a postgraduate training college). Today there are three Chairs of geography at Copenhagen University, two at Aarhus, and one at Danmarks Laererhøjskole.

Libraries attached to geographical research

Several Danish libraries hold geographical literature, the two principal libraries in Copenhagen are The Royal Library, which is the main library for human geography and maps, and the University Library, 2nd section, which provides literature on physical geography. Further, mention should be made of the library of The Royal Veterinary and Agricultural High School (Den kgl. Veterinaer- og Landbohøjskoles Bibliotek) with its collection of agricultural-geographical literature; the library of the Statistical Department (Statistisk Departements Bibliotek); the library of the Copenhagen School of Economics & Business (Handelshøjskolens Bibliotek), and Denmark's Technical Library (Danmarks tekniske Bibliotek), in which is found literature of importance to the study of commercial geography. Finally, there is the Geographical Institute of Copenhagen, which has a large library of literature covering all aspects of the geographical research, and, besides, a fine collection of wood engravings. In Aarhus, Jutland, The State Library is the main library for geography.

The University Library of Copenhagen

The first important library devoted to geographical research was the University Library, which was founded in 1482 and housed above the vaults of the Trinity Church at the Round Tower of Copenhagen. During the Big Fire of Copenhagen in 1728 the church was burnt down and the library almost completely destroyed. After the rebuilding of the church it was, however, reestablished, and in 1861 the library was moved to a new building near the university. In 1927 the University Library became Denmark's main library for the natural sciences and medicine, and in 1938 the literature on these two disciplines was moved to a building erected in the years 1935-1937. The name and address of this library is, The University Library, 2nd section, 49 Nørre allé, DK-2200 Copenhagen N. - At a later date the university sections of the natural sciences and medicine were built near this library. - For literature on physical geography it would be natural to apply first to the University Library, 2nd section, because they have a union catalogue of the literatures of most specialized university libraries, from many of which material is only available through the University Library, 2nd section

The Royal Library (Det kongelige Bibliotek), 8 Christians Brygge, DK-1219 Copenhagen K.

As mentioned above, The Royal Library is the main library for human geography and maps. It is Denmark's national library and the state library of the humanities and sciences. It was founded in 1657 by Frederik III and housed in the Castle of Copenhagen up to 1673 when it got its own building. There it remained until 1906. In 1697 The Royal Library became copyright deposit library for all that is printed in Denmark. On November 15, 1793, it became a public library. Till then it had only been accessible to the court and a few scientists. In 1906 the library was moved to its present building. The total stock of books is now approx. two millions, one third of which belongs to the Danish Department.

The Geographical Library (Det Kongelige Danske Geografiske Selskabs Bibliotek), 68 Haraldsgade, DK-2100 Copenhagen Ø.

The first geographical society in the world was formed in 1821 in Paris, viz. the "Société de Géographie". One of the founders was the Dane, Malthe Conrad Bruun, 1775-1826.

He had been exiled from Denmark for political reasons and settled down in France where he published his epochmaking work, "Précis de la Géographie Universelle, 1810-29". From then on, such societies were formed all over Europe, in 1876 The Royal Geographical Society was founded in Copenhagen. This society created the basis of a geographical library as its publication, the 'Geographical Periodical', became an exchange object for periodicals of other geographical societies in Europe; very soon this resulted in a considerable wealth of material and a general lack of space. Owing to the kind co-operation of the Topographical Department of the Ordnance Survey, a room in the top storey of the Victualling Yard (Proviantgården) was made available to the society in 1877. This room was, however, unfit for lending service and in order to make up for this deficiency professor O. Olufsen offered to lend a room of his private apartment. In 1907 the society moved to 135 Gothersgade and in 1909 it ventured to open a library with reading room and conversation room in one of the finest buildings of Copenhagen, at the corner of Kongens Nytorv (the King's New Square) in the centre of town In 1923 the society moved to 27 Amaliegade, where it was housed till 1933. The subsequent address of the library was The National Museum where it was closely connected to the Ethnographical Collections. In 1956 the library was transferred to the geographical laboratory. It is still the property of the Geographical Society, but is being used as part of the Geographical Institute. All new book acquisitions are available in the reading room to scientists and students, to the members of the society and, as a matter of fact, to all who need it. The Geographical Society receives about 500 periodicals a year, not to mention monographs, etc.

The State Library in Aarhus (Statsbiblioteket i Aarhus), Universitetsparken, DK-8000 Aarhus C.

The State Library is the main library of physiography and human geography to the Geographical Institute in Aarhus. The State Library was founded in 1902, its holdings consisting chiefly of duplicates of Danish literature from The Royal Library and, in addition, several considerable Danish private collections. Since 1902 it has been deposit library for all Danish literature printed. Concerning foreign literature, it is the duty of The State Library in its capacity of university library to provide all literature necessary in the university work, and in its capacity of research library to acquire literature of interest to the borrowers of the state-supported library.

Libraries attached to cartographical research

The Maps and Prints Department of the Royal Library
(Det kongelige Biblioteks Kort- & Billedafdeling)

The Royal Library is in possession of Denmark's largest collection of maps and ever since its foundation in the 1670s atlases and maps have been deposited here. It was not until 1902 that it was decided to establish a separate maps and prints department. The nucleus of the collection of maps and atlases was primarily the library's own holdings. The maps were not catalogued, however, so it is easy to imagine how difficult it must have been for researchers to find out whether the particular items of interest were really there. Among the atlases in the library collection two are of special value as regards quality and quantity, viz. Frederik the Fifth's Atlas and Atlas Major. Both contain many unique maps. Frederik the Fifth's Atlas was compiled during the reign of Frederik V, 1746-66, and consisted of maps and prospects in loose leaves prints as well as drawings owned by the King. Once within bindings the material made an atlas of 55 volumes with a total of 3,535 sheets. - Atlas Major is very much like Frederik the Fifth's Atlas, but contains fewer drawings. When the Maps and Prints Department was established, several other atlases were transferred from the magazines of The Royal Library, and all important cartographers are now represented through one or several works.

From the very beginning the collection was large and representative and soon it became the largest collection of maps in Denmark, partly through acquisition, partly through transfers of maps and atlases from other institutions. Also gifts contributed to the growth of the collection. Thus, in 1911, the consul I.F. Erichsen presented The Royal Library with about 20 000 sheets, comprising maps, topographical sheets, portraits, etc., everything in excellent condition.

In 1910-11 the department received about 1 200 maps and drawings from the collection of the Corps of Royal Engineers, mostly consisting of fortifications, and among these many foreign ones. They are all maps drawn by hand and many of them of extremely high historical value, several scientists have found them to be unique source material.

The Hydrographic Department (Søkortarkivet) has also contributed with valuable maps of paramount importance to the exploration history of the North Atlantic area, e.g. Hans Resen's original map from 1605. Resen's map was given by the Hydrographic Department in 1908 together with 5 000 other maps. The same institution has later presented The Royal Library with a great number of important maps.

In 1918 the library received from the "Committee for the Denmark Expedition" the maps that claimed so many victims, the original sketch maps of the north-eastern parts of Greenland. The maps were drawn by Høeg Hagen in 1907 during the sledge expedition with Mylius Erichsen and Jørgen Brønlund and completed the mapping previously commenced from the west. The members of the expedition never got back to their ship, but perished. The sketch maps were found near the body of Jørgen Brønlund.

In 1976 the department received 60 000 recent maps and 2 000 historical maps from the Geodetic Institute and an agreement was made with said Institute to cooperate in future by means of an exchange system.

Literature: Ib Rønne Kejlbo, "Manuscript Maps in the Frederik den Femtes Atlas", Copenhagen 1969. Leif Makholm, "Danmarkskort publiceret før 1700", København 1975. Preben Sonne Jørgensen, "Fortegnelse over kort i samlingerne på Det kongelige Bibliotek, København, der kan tjene Landbohistoriske studier af lokaliteter på Fyn, Langeland og omliggende øer indtil midten af det 19. århundrede", København 1974.

Several other institutions and private persons have donated maps and atlases, thus contributing to the growth of the holdings of the Maps and Prints department. The collections are still being increased through legal deposit, gifts, and acquisitions, and the field of acquisition comprises so to speak all categories of cartographical material.

The maps were previously arranged alphabetically, but about ten years ago the department began to use a different method of cataloguing, a kind of decimal system according to which each catalogue card has a location mark, e.g. 1111,2-5-1968. The location mark consists of three groups: the first group (1111,2) indicates the region, for instance a state. The second group (5) indicates the subject or theme, in this particular case it shows that it is a geological map. The third group (1968) refers to the year in which the map was made.

In addition to the maps in the Maps and Prints Department another collection of maps is kept by the Manuscript Department of The Royal Library.

The Map Collection of The National Archives (Rigsarkivets Kortsamling), 9 Rigsdagsgården, DK-1218 Copenhagen K.

The map collection of the National Archives is mainly based on transfers from various institutions. The map material for Southern Jutland and Holstein is particularly comprehensive, and the maps of these areas principally come from the Exchequer. In the years 1822 and 1823 the Customs Department (Generaltoldkammeret) and the Harbour Boards (Havnekommissionerne) contributed maps of Danish towns and harbours, and the Colonial Office maps of former Danish colonies. Deposits from the Ministry of War mainly consist of maps of fortifications. These maps are kept in a "map collection" to which belongs a card index. The maps are arranged according to the provenance principle. Apart from the maps in this "map collection" there are, of course, maps which are enclosures to many of the documents kept in the National Archives. Few of them are registered, so the maps must be sought among the documents. This, unfortunately, makes the search rather difficult.

The Map Collection of The Land Registry, The Archives of The Land Registry (Matrikeldirektoratets Arkiv), 13 Titangade, DK-2200 Copenhagen N.

The map collection of the Land Registry contains material of great importance to the topographical research. The first registration with mapping was begun in 1806-07 and already existing map material was included in the registration. The major part of it dated back to the Enclosure Movement, during which maps in two copies had been made up in accordance with a government circular of 1781. One copy was kept at the Exchequer, and from there transferred to the archives of the Land Registry. Cadastral maps of Danish towns were not prepared until 1863, however. Apart from the maps directly connected to the registration, the Land Registry has a collection of special maps of railways and roads.

The Chart Collection of The Royal Hydrographic Department (Det kongelige Søkortarkivs Kortsamling), 19 Esplanaden, DK-1263 Copenhagen K.)

The Royal Hydrographic Department, which is a map printing institution, also keeps a chart collection. A printed catalogue from 1895 registers the old part of this collection, "The antique chart collection of the Royal Hydrographic Department" 1895. It should be added, however, that a considerable number of the charts mentioned in the catalogue have been transferred to the National Archives and The Royal Library.

Literature: "Søkaart-Archivets Antike Kaartsamling", Kjøbenhavn 1895.

The Map Collection of The Geodetic Institute (Geodaetisk Instituts Kortsamling), Proviantgården, 7 Rigsdagsgården, DK-1218 Copenhagen K.

This institution, which is in charge of the Danish topographical surveying and the publication of Danish topographical maps, has a map collection containing the original maps and copper plates for the map printing of The Royal Danish Academy of Sciences and Letters and also various historical as well as modern maps acquired through exchange with foreign map printing institutions, see the last paragraph of the article about The Maps and Prints Department of The Royal Library.

The Private Library of Her Majesty The Queen of Denmark, formerly: of His Majesty The King of Denmark (Hendes Majestaet Dronningens Håndbibliotek), Christiansborg, DK-1218 Copenhagen K.

This collection is not open to the public, but special permission can be obtained by scientists. The private library of Her Majesty the Queen of Denmark possesses a collection of maps and atlases. Among these atlases there is one of particular value, the "Juliane Marie Atlas", which formerly belonged to Queen Juliane Marie, 1729-96. The queen, a daughter of the Duke Ferdinand of Braunschweig-Wolfenbüttel, was married to the Danish King Frederik V in 1752. The "Juliane Marie Atlas" consists of 37 volumes, bound in the same way as the "Frederik the Fifth's Atlas" in The Royal Library.

Literature: H.F. Kiaer; "The 'Juliane Marie Atlas' in Copenhagen". In: Imago Mundi. Vol. XX, 1966, p.82-84.

The Map Collections of The Military Archives

The Military Archives keep an abundant supply of map material of great importance to research, such as excellent maps of several towns which have had fortifications, as well as maps of manoeuvre and maps of war, many of which carry useful topographical information. Moreover, the maps bear witness that many of the officers were fine cartographers.

The following military archives should be mentioned:

THE ARMY ARCHIVES (Haerens Arkiv), 4 Slotholmsgade, DK-1216 Copenhagen K.

In the Army Archives there are about 10 000 maps, not only of Danish areas, but also of several other European localities. A card index of the maps is available.

THE DRAWING ARCHIVES OF THE ARTILLERY (Artilleriets Tegnearkiv), MUSEUM OF ARMS AND UNIFORMS (Tøjhusmuseet), 3 Tøjhusgade, DK-1214 Copenhagen K.

The Drawing Archives of the Artillery hold a series of plans of Norwegian-Southern Swedish fortifications from the second half of the 17th century. Most of them are copies, but several give useful information about buildings, street names, etc. A card index is available.

THE NAVY LIBRARY (Marinens Bibliotek), 62 B Overgaden oven Vandet, DK-1414 Copenhagen K.

The Navy Library has several excellent charts all of which are catalogued.

The State Library in Aarhus (Statsbiblioteket).

The only large general map collection outside Copenhagen is kept in The State Library of Aarhus. It was established in 1902 and mainly based on duplicates from The Royal Library in Copenhagen. The State Library of Aarhus moved in 1962, and only then did the map collection find proper conditions; the collection is now being rearranged and catalogued. The map collection in Aarhus also receives copies of all maps being printed through legal deposit.

Apart from the map collections mentioned above there is map material in the museums and archives of most Danish towns. The major part of the maps contained in such collections are of local interest, however, and primarily of interest to people who want topographical material about a particular Danish town and its surroundings.

LES COLLECTIONS DU DÉPARTEMENT DES CARTES ET PLANS DE LA BIBLIOTHÈQUE NATIONALE DE PARIS

Par Edmond Pognon

Selon les plus récentes statistiques, le Département des Cartes et plans de la Bibliothèque nationale de Paris conserve environ 1 500 000 feuilles de cartes géographiques et de plans, auxquelles il faut ajouter quelques 5 000 atlas et 40 000 volumes d'ouvrages intéressant l'histoire et les techniques de cartographie, plus 90 collections de périodiques vivants. Il héberge en outre la Bibliothèque de la Société de Géographie de Paris.

Comme les autres départements de la Bibliothèque nationale, il est à la fois un conservatoire de documents anciens, donc intéressants surtout pour l'historien, et un centre de documentation actuel, pouvant être utile à toutes sortes d'activités d'aujourd'hui. Il conserve en effet tout ce qui est entré, depuis le temps de Louis XIV jusqu'à nos jours, à la Bibliothèque royale, puis nationale, en fait de documents cartographiques.

Origines du fonds

Quand, en 1828, fut préparée une grande ordonnance de réformation de la Bibliothèque royale, un savant de tout premier ordre obtint qu'y fut crée un département des collections géographiques. C'était Edme-François Jomard, membre de la première promotion de l'Ecole Polytechnique, qui avait fait partie de la mission scientifique emmenée par Bonaparte en Egypte, avait ensuite pourvu à la monumentale publication des documents de toute sorte rapportés des bords du Nil et avait été, en 1821, un des principaux fondateurs de la Société de Géographie de Paris.

La Bibliothèque avait jusqu'alors conservé au Cabinet des Estampes les cartes entrées dans les collections royales. C'étaient d'abord celles que déposaient, depuis l'institution du "Privilège du Roi", les éditeurs français soucieux de se protéger de la contrefaçon: on avait donc là, pratiquement, toute la production cartographique du royaume depuis les origines. C'étaient en outre celles qui avaient pu se trouver dans des collections privées acquises par le roi, dont la plus importante était celle de Roger de Gaignières, entrée vers 1700. Il y avait là beaucoup de cartes de provenance étrangère, italiennes du XVIe siècle, néerlandaises plus récentes, allemandes, anglaises. S'y joignirent celles que les gardes de la Bibliothèque royale eurent le bon esprit d'acquérir dans le cours du XVIIIe siècle. Une collection particulière beaucoup moins considérable, mais très intéressante, celle de Joseph Nicolas Delisle, était entrée en 1754: elle était essentiellement composée de cartes de Russie, accompagnées de mémoires manuscrits.

Un autre très important ensamble était la collection réunie à la fin du XVIIe siècle par Nicolas de Tralage, léguée par lui à l'abbaye de Saint-Victor de Paris et devenue bien national par décision du gouvernement révolutionnaire. Toute la production des XVIe et XVIIe siècles y était représentée, si largement qu'on peut presque la dire complète, surtout pour les cartes non françaises.

Ayant pris en charge ces deux grandes collections, le nouveau département devenait d'emblée un conservatoire pratiquement exhaustif de tout ce qui avait pu s'imprimer comme cartes et comme plans en Europe depuis la fin du XVe siècle.

Quelques années plus tard, le département s'enrichit d'une partie des atlas anciens et récents jusqu'alors conservés au département des Imprimés.

Désormais, par le jeu du dépôt légal pour les cartes françaises, par acquisitions et échanges pour les cartes étrangères, le département s'est tenu à jour par rapport à la production mondiale. Il a en outre complété en tant que de besoin son fonds ancien par des acquisitions exceptionnelles dont la plus prestigieuse, en ces dernières années, l'a enrichi du plus beau des quatre exemplaires connus de la carte d'Europe gravés en 1564 par le Vénitien Giovanni Baptista Contarini.

Les cartes nautiques "sur vêlin"

Mais les cartes imprimées ne sont pas toute la cartographie. Jomard connaissait l'importance des cartes exécutées à la main, et surtout des plus anciennes. Le recueil de fac-similés malheureusement inachevé qu'il avait intitulé *Monuments de la Géographie* en témoigne. Et - imité en cela par ses successeurs jusqu'à nos jours - il a pris à tâche de faire entrer dans le département le plus grand nombre possible de ces rares "monuments". On sait que les plus précieux et les plus intéressants sont ceux qu'on appelle couramment, encore qu'improprement, "portulans". Ce sont des cartes nautiques, exécutées sur parchemin ou sur vélin, dont les plus anciennes ne figurent que le bassin méditerranéen mais qui ont élargi leur assiette à mesure des découvertes. Le département s'honore de conserver la première en date de toutes, la fameuse "carte Pisane" qui remonte aux environs de 1300. C'est Jomard qui l'a acquise en 1839. Celle de Mecia de Viladeste, qui est de 1413, la carte dite de Christophe Colomb - attribution fortement contestée - , celles qui forment l'atlas nautique de Lopo Homem (1519) s'y joignirent, sans parler de bien d'autres moins célèbres. Aujourd'hui le fonds des "cartes nautiques sur vélin" en compte 183, dont le catalogue, par Myriem Foncin, Monique de La Roncière et Marcel Destombes, a été publié en 1964. Ce nombre tient compte, outre celles du département, de celles de la Société de Géographie et de celles qui appartiennent au Service hydrographique et océanographique de la Marine, mises en dépôt au département depuis 1940.

De ces cartes nautiques dites "sur vélin" - bien qu'elles ne soient pas toutes exécutées sur un support aussi luxueux - il y a lieu de rapprocher les 88 cartes sur parchemin dites "géographiques", dont la provenance et la nature sont extrêmement diverses, et dont le catalogue, dû à Roger Hervé, Edmond Pognon et Henri Hugonnard-Roche, a été publié en 1974.

Cet ensemble de cartes à la main de haute époque, qui met le département au premier rang dans le monde, ne peut être négligé par aucun historien de la cartographie ancienne. Les bibliographies qui se lisent à la fin des deux catalogues témoignent d'ailleurs des innombrables travaux pour lesquels elles ont déjà été mises à contribution, par des savants des pays les plus divers.

Entrées de collections particulières

Il n'est pas sans intérêt de mentionner quelques collections privées qui sont entrées au département depuis sa création.

Ce furent d'abord, dès 1832, les 1613 feuilles et les 34 volumes légués par Klaproth, qui concernent surtout les pays d'Asie; l'année suivante, les deux portefeuilles et la grande carte roulée achetée à la vente Rémusat, où on trouve des cartes historiques, certaines dues à Guillaume Delisle et à Buache, et des cartes de pays de l'Extrême-Orient. En 1834, c'est l'entrée de la collection Paccaroni, qui comporte 70 cartes à la main dont une sur vélin, et 118 cartes vénitiennes du XVIe siècle. La collection du chevalier Hennin, vendue en 1836, fait entrer un atlas italien du XVIe siècle, réalisé par Gastaldi, et plusieurs cartes de la même époque, dont certaines gravées sur bois. Quatre ans plus tard arrivent, par achat à Fleury Hérard, trois plans-relief - dont un de l'Acropole d'Athènes - exécutés sous le 1er Empire par le diplomate Fauvel, de qui proviennent aussi plusieurs cartes de Barbié du Bocage et de son maître, d'Anville. Six portefeuilles contenant, parmi leurs 1227 pièces, de nombreuses cartes historiques, appartenant à la collection Gosselin, sont achetés en 1842 au fameux libraire de Bure. La même année, 100 feuilles à la main figurant les côtes de la Manche, provenant de la famille d'Orléans, viennent enrichir le département en cartes hydrographiques. En 1844, nouvel apport de cartes ayant appartenu à Hennin: 218 cartes gravées diverses, 139 manuscrites, 46 dressées par des militaires en campagne, 83 plans de ville.

Les canaux et les rivières de France, autrement dit le réseau de navigation fluviale, font leur entrée en 1850 avec 596 pièces provenant de la collection du Jurassien Cordier. L'année suivante, c'est le tour des possessions françaises de l'Atlantique, Guyane et Antilles surtout, avec 63 cartes dont beaucoup sont manuscrites. Elles avaient appartenu à la collection Lefèvre, comme celles qui devaient les rejoindre en 1855, et où se retrouvent les Antilles avec la Guadeloupe et l'Allemagne. Les fortifications, déjà largement représentées depuis les guerres des derniers Bourbons, s'enrichissent encore en 1857 avec des cartes provenant de Danquin et qui concernent surtout la Flandre et l'Allemagne.

Tous ces enrichissements sont dus à l'active vigilance de Jomard, qui meurt en fonctions en 1860. Dès 1862, son successeur obtient des Archives nationales la cession d'un très grand nombre de cartes. En 1866, la vente de la collection De Manne permet d'acquérir 992 feuilles, dont 129 dues pour la plupart à la main de d'Anville. Et en 1871, ce sont 63 cartes manuscrites et calques dus à un autre illustre géographe, Barbié du Bocage, qui entrent, achetées à Lefèvre

Grandes collections entrées au XXe siècle

Le XXe siècle a vu se poursuivre, parallèlement à l'entrée régulière de la production française et étrangère, l'accroissement du fonds ancien du département. Plutôt que les petits achats au jour le jour qui se poursuivent, mais qui deviennent moins intéressants parce que les belles collections particulières se font plus rares en France, il faut évoquer plusieurs entrées massives, que le chercheur doit connaître.

Ce fut d'abord la remise par le ministère des Affaires étrangères de la collection d'Anville. Jean-Baptiste Bourguignon d'Anville (1697-1782) avait été le plus éminent des "géographes du Roi". Tout en poursuivant une oeuvre immense de cartographe, il avait, à son usage, réuni plus de 10 000 cartes. Louis XV les lui acheta et les mit, très judicieusement, à la disposition de ses diplomates. La collection resta au ministère des Affaires étrangères sans être tenue à jour, et par conséquent de moins en moins utile. En 1924, Charles du Bus, conservateur de la "Section de Géographie", sut faire comprendre aux amis qu'il avait au Quai d'Orsay l'intérêt qu'il y aurait à transférer cette documentation, désormais purement historique, à une maison où elle pourrait plus aisément être consultée par les chercheurs. Ainsi fut fait. Ce magnifique ensemble de cartes classées selon un ordre géographique, donc très commode, qui couvre le monde entier et dont les pièces s'échelonnent du XVIe siècle au troisième quart du XVIIIe, est aujourd'hui un des plus beaux fleurons de la couronne du département. Il en existe un catalogue pièce par pièce, manuscrit, dû à Barbié du Bocage, et un état sommaire qui lui sert de très utile table des matières, publié par Charles du Bus dans le *Bulletin de la Société de Géographie* de 1926.

Le Service hydrographique et océanographique de la Marine conservait tous les travaux cartographiques exécutés depuis le temps de Colbert par les ingénieurs hydrographes français. Minutes et calques, remontant jusqu'au XVIIe siècle, couvraient à peu près l'ensemble des côtes de la Terre entière. Quelques cartes nautiques de haute époque, dont l'origine est mal connue, s'y joignaient; certaines d'entre elles, on l'a vu, furent confiées dès 1940 au département. Mais c'est en 1947 qu'y furent mises en dépôt toutes les pièces antérieures à 1800. Puis, en 1962, lui furent données les cartes gravées au Service hydrographique au cours des XIXe et XXe siècle. Elles servirent à compléter les séries de ce genre déjà conservées par le département. Mais le fonds ancien, qui reste la propriété du Service hydrographique, est soigneusement maintenu distinct; il n'en est pas moins géré par le département dans les mêmes conditions que ses collections propres. Le chercheur peut les consulter et en obtenir des photos sans s'apercevoir de la différence. L'intérêt historique de ces cartes de côtes, qui donnent les états successifs de configurations particulièrement instables de la surface terrestre, n'a pas besoin d'être souligné. L'inventaire dactylographié, pièce par pièce, suivant l'ordre géographique qui est celui du rangement dans les portefeuilles, permet d'y trouver aisément celles qu'on y cherche.

Les 2077 cartes et plans de villes, dont 99 pièces exécutées à la main, qui remplissent les 18 beaux atlas reliés en veau de la collection de Gaston d'Orléans n'ont pas eu beaucoup de chemin à faire pour entrer en 1962 au département des Cartes et plans: elles étaient dans la Bibliothèque du roi depuis 1660, date à laquelle le jeune Louis XIV en hérita de son oncle, le turbulent Gaston d'Orléans, frère de Louis XIII. Pendant trois siècles, elles étaient restées aux Imprimés. Le catalogue est en cours de rédaction par les soins de Françoise Lapadu-Hargues. Il s'agit d'un très bel ensemble de cartes de la première moitié du XVIIe siècle, couvrant toute l'Europe, et en magnifique condition.

Atlas anciens

Après avoir énuméré tant de cartes, peut-être est-il bon de rappeler la présence au département des atlas. Tous ceux de la période contemporaine y sont représentés, prélevés sur le dépôt légal ou acquis à mesure de leur publication. Quant à la collection des atlas anciens, elle peut, elle aussi, être dite complète, qu'il s'agisse des tout premiers, publiés à Rome et à Venise au XVIe siècle, aussi bien que des recueils de Mercator, d'Ortelius, de Van Keulen et des autres cartographes néerlandais ou allemands, sans omettre, bien entendu, les Anglais et les Français. Les recueils de plans de villes, notamment les fameuses *Civitates Orbis Terrarum* de Braun et Hogenberg, ne manquent pas davantage à l'appel.

Globes et instruments astronomiques

Ce n'est pas encore tout. En même temps que ces innombrables feuilles et volumes, le département conserve en effet une quantité non négligeable d'objets en rapport avec la géographie, et aussi avec l'astronomie.

Ce sont d'abord des globes terrestres, dont certains comptent parmi les plus anciens connus: le fameux *Globe Doré*, qu'on date d'environ 1527 et qui n'a pas fini de livrer tous ses secrets; *le Globe vert, le Globe de Bois, le Globe de Rouen, le Globe Lécuy*, tous exécutés dans la première moitié du XVIe siècle, ou très peu après; un fac-similé, commandé par Jomard, du fameux globe exécuté en 1492 par Martin Behaim, dont l'original est conservé à Vienne. Du XVIIe siècle, plusieurs paires de globes - un terrestre et un céleste -, dont les plus prestigieux sont ceux de Vicenzo Coronelli, d'environ 1 mètre de diamètre, réductions gravées des deux énormes sphères que le célèbre Vénitien exécuta pour Louis XIV à la demande du cardinal d'Estrées et qui, après avoir été installés par le roi à Marly, occupèrent lamentablement entreposés dans l'Orangerie de Versailles, où ils demeurent invisibles. Des paires du XVIIIe siècle, dues à Bion, à Doppelmaier entre autres, deux globes célestes arabes, dont un du XIIe siècle, complètent cette précieuse collection, dont l'énumération, très sommaire peut se lire dans la *Liste des globes terrestres et célestes anciens conservés dans les collections publiques de France*, publiée en 1970 sous l'égide du Centre national de la Recherche scientifique, et dont le mérite revient à Mme Gabrielle Duprat, anvien conservateur en chef de la Bibliothèque du Muséum national d'Histoire naturelle.

A ces globes s'ajoutent trois remarquables sphères armillaires en laiton. Deux sont montées sur un simple pied de bois noirci; elles paraissent remonter au XVIIe siècle. L'une représente l'univers selon Ptolémée - avec la Terre au centre -, l'autre est héliocentrique, conforme au système de Copernic. La troisième, géocentrique, et conçue pour expliquer le phénomène des éclipses, est animée par un mouvement d'horlogerie et repose sur un magnifique trépied en bois sculpté et doré, d'époque Régence; elle a pour auteur Jérôme Martinot.

Non moins luxueux sont deux calendriers astronomiques - ou mieux astrologiques -, constitués par de grands cadrans hexagonaux en laiton, munis d'un mécanisme intérieur, montés sur des trépieds en bronze doré figurant des dauphins, et couronnés de l'emblême solaire de Louis XIV, également en bronze doré. Cette décoration indique clairement qu'ils furent exécutés pour le Grand Dauphin, fils du Roi Soleil, donc antérieurs à 1711, date de la mort de ce prince. Ils sont signés par Thuret.

Plusieurs astrolabes complètent cette collection d'instruments astronomiques. Le plus aucien est aussi le plus ancien connu. Il est l'oeuvre de Ahmed Ibn Khalaf et date du IXe siècle.

Le département conserve encore d'autres instruments, parmi lesquels se remarquent deux boussoles chinoises du XVIIIe siècle.

Classement et catalogues

Le rangement matériel des cartes et leur conservation posent des problèmes difficiles que connaissent bien tous les responsables de cartothèques. L'idéal serait évidemment de pouvoir les ranger selon un ordre géographique. C'est possible dans le cas d'un fonds qui ne s'accroît pas. Par exemple, la collection d'Anville est, nous l'avons dit, classée de cette manière. Mais le fonds général, auquel s'ajoutent en moyenne 9 000 feuilles par an, ne pourraît l'être qu'au prix de "mouvements" continuels et parfois de grande ampleur, qu'il serait illusoire d'envisager. Il faut donc se résigner à ranger les cartes, par format, dans l'ordre où elles arrivent. On réserve de la place pour les cartes à grande échelle composées de plusieurs centaines de feuilles, dont la publication s'échelonne sur des années, et qui doivent, naturellement, être groupées ensemble.

Les catalogues permettent de retrouver la carte désirée, dont la cote est constituée par le numéro qui indique sa place dans le format où elle a été rangée. Ils sont sur fiches, rédigées selon des règles si judicieuses qu'elles ont été retenues, presque telles quelles, pour l'I.S.B.D. (CM), c'est-à-dire "International Standard Bibliographic Description (Cartographical Material)". Il y a un catalogue alphabétique par nom géographique, un catalogue par nom d'auteur, un catalogue analytique, dont les vedettes sont constituées par un mot qui désigne le thème particulier de la carte. Il serait souhaitable de publier ces catalogues, par exemple en reproduisant en offset les fiches juxtaposées de manière à former des pages; ce sera peut-être l'oeuvre de Monique Pelletier, nouveau conservateur en chef du département, pourvu que les moyens financiers lui en soient accordés ...

La Bibliothèque de la Société de Géographie

Le chercheur qui vient se documenter au département des Cartes et plans dispose, en plus du fonds qui vient d'être amplement décrit et des ouvrages de référence et périodiques spécialisés signalés dès la première page de cet article, des 400 000 volumes de la Bibliothèque de la Société de Géographie, qu'il héberge depuis 1942. Cette bibliothèque, qui demeure autonome, s'enrichit régulièrement de volumes reçus en service de presse grâce au supplement bibliographique de la revue de la Société, *Acta geographica*, et de périodiques étrangers qu'elle reçoit en échange du service de cette même revue. Mais elle possède aussi un très beau fonds de livres géographiques anciens, dont un grand nombre lui est échu avec la bibliothèque du prince Roland Bonaparte, où se remarque entre autres un splendide exemplaire de la *Chronique de Nuremberg*. Rappelons aussi qu'elle a apporté au département plusieurs cartes nautiques sur vélin, notament une belle carte de la Méditerranée par Giaccopo de Maggiolo (1976), et une très curieuse vue de la prise de Macassar par les Hollandais en 1660.

Cette énumération, peut-être encore incomplète, des richesses du département des Cartes et Plans de la Bibliothèque nationale le fait assurément apparaître comme un des deux ou trois plus importants du monde. Mais c'est aussi la diversité de ses collections qui frappe: il conserve, comme un dépôt d'imprimés, des livres; comme un cabinet d'estampes, les pièces gravées que sont les cartes, et parmi lesquelles certains plans de villes relèvent plus de l'art figuratif que de la cartographie; comme un département des manuscrits, les feuilles enluminées d'or et de couleurs vives que sont le plus souvent les cartes nautiques sur vélin; comme un musée, les objets d'art, parfois somptueux, on l'a vu, que sont ses globes, ses sphères armillaires, ses astrolabes. C'est un lieu voué, certes, à la recherche et aux études savantes, mais aussi offert à la contemplation esthétique, et au rêve ...

COLLECTION OF CARTOGRAPHIC WORKS IN THE LENIN STATE LIBRARY OF THE USSR

By Anna V. Kozlowa

The Lenin State Library of the USSR has one of the country's richest collections of home and foreign cartographic works amounting to some 190 000 items which are of scientific, economic and cultural importance.

The history of building up the Library's cartographic collection is inseparably connected with the process of shaping its other stock. For instance, apart from books, journals and other materials, collections of N.P. Rumyantsev, A.S. Norov, D.A. Guryov and others that made up the bulk of the stocks of the Moscow Public Library, included a certain number of geographic maps, plans and atlases. N.P. Rumyantsev's collection alone numbered about 500 such items. The process of enriching the Library with maps and atlases was uninterrupted.

The Library received obligatory copies of domestic maps and atlases on the same basis as books. However, there was no absolute guarantee that the Library had received obligatory copies of maps and atlases since 1862, as the "Regulations on the Moscow Public and Rumyantsev Museum" adopted in 1862 had no provision to the effect that all the cartographic works published in the country should be sent to the Library.

Essential classifications concerning the obligatory copy, which were very important for the supply of the Library with maps, were introduced in the Statutes of the Moscow Public and Rumyantsev Museum, which read that the "Museum is granted the right to receive an unpaid copy of every item printed, engraved and lithographed both by private persons and government departments in Russia ... Copies of books ... as well as copies of sheets ... geographical maps, plans ... shall be sent to the Museum".[1)]

An important source of enriching the Library with maps and atlases in pre-revolutionary times were donations made by prominent Russian geographers B.S. Bodnarsky, and A.A. Borzov, the well-known traveller N.N. Miklukho-Maklai, historian P.Ya. Chaadayev, sinologist K.A. Skachkov, honorary corespondent of the Moscow Public and Rumyantsev museum A.N. Neustruyev and others.

During the first decades of the Library's existence the building up of the domestic stock of its maps and atlases depended to a considerable extent on the transfer of duplicates from the Imperial Public Library which had received an obligatory copy since 1810 already. Certain maps and atlases were received from state institutions of Russia. Because of lack of funds very few foreign maps and atlases were acquired.

During the first years of the existence of the Library of the Moscow Public Rumyantsev Museum cartographic publications were either included in the thematic collections of the museum or placed in the Library among books and other items of printed matter. The reason for placing maps and atlases in this or that location of the book repository was the appearance and format of the publication.

Maps put in cardbook cases and atlases in the form of books and albums were placed on shelves alternately with books and journals. Only the maps and plans printed on sheets were placed on separate shelves.

Lack of spare room in the book repository had for many years prevented the organization of a specialized cartographic stock despite the fact that the volume of cartographic works stored in the Library has considerably increased since 1921, as the Library has received a second obligatory copy of all items of printed matter. Already in 1925 the Library acquired about 900 maps.[2)]

The stock of maps was established shortly before the Great Patriotic War when a new book repository was built, and it was possible to divide the Library stock by type of publication.

A considerable number of maps issued in cases and boxes, and all the atlases printed before the 1950s had still been kept among book collections.

During the Great Patriotic War the entire cartographic stock was preserved. As for the enrichment of the stock, it was actually interrupted because of the change in the nature and direction of cartographic-publication activities which were fully subordinated to the interests of the country's defense.

After the Great Patriotic War, in connection with the increased issue of domestic cartographic works and due to the fact that the Library received a third obligatory copy, the number of maps and atlases received by the Library had substantially gone up.

The composition of the cartographic stock of the Lenin State Library is universal; it includes domestic and foreign (in many languages of the world) one-sheet and multi-sheet maps, plans and map-schemes, one-volume and multi-volume atlases and relief maps which are different in subject matter, purpose and polygraphic execution, as well as explanatory notes, indexes and other textual, graphic and illustrated supplements to cartographic editions.

The Library has collected a complete set of Soviet maps and atlases, a rather large stock of domestic cartographic works issued before 1917 and a large stock of foreign maps and atlases which is of scientific, practical and historical value. Soviet editions of maps and atlases make up 52 per cent of the stock, domestic editions issued before 1917 - 15 per cent and foreign editions - 33 per cent.

Soviet maps and atlases

The Library's collection of Soviet cartographic editions is the largest in the country. Its creation has been facilitated by the successes of the Soviet cartographic activities launched under the Decree of the Council of People's Commissars of the RSFSR signed by V.I. Lenin in 1919 on the establishment of the Higher geodesic department of the Republic. [3)] The department was charged with the preparation and edition of different-scale maps of state importance, largely for the national economy and defence, and with the renewal of the issue of geographical maps intended for the public at large.

Large-scale and varied cartographic output of the Soviet-Union directly affected the nature of the Library's cartographic stock.

The stock includes fundamental scientific-reference cartographic works: general geographic, special and complex atlases, sectoral thematic maps and atlases, reference geographic and political-administrative maps, as well as educational and tourist cartographic works.

Of the greatest value are large general geographic and special fundamental atlases reflecting the achievements of Soviet science. These include the Great Soviet Atlas of the World, the 3-volume Marine Atlas, the general geographic Atlas of the World, the complex Atlas of the USSR, the physico-geographic atlas of the world, the Atlas of the Antarctic, the Atlas of the history of geographic discoveries and studies, the Atlas of the peoples of the world, the Atlas of the development of the USSR's national economy and culture, the Atlas of the agriculture of the USSR, and others.

The collection of Soviet fundamental atlases is the best in the world. Due to their scientific content and great informativeness, many of these atlases have made up the golden fund not only of the domestic but of the world geography and cartography as well. For instance, the Atlas of the Antarctic is the first complex cartographic work which gives a comprehensive coverage of the history of exploration and of the nature of the sixth continent and the ocean surrounding it. As one of the best scientific works the Atlas has been awarded the State Prize. The Marine Atlas has also been awarded the State Prize.The first volumeof the Great Soviet World Atlas won the first prize (Grand Prix) at the International Paris Fair; the physico-geographic world atlas won a 1st degree diploma at the All-Union competition for the best design and polygraphic execution of the book; a number of atlases were awarded diplomas and medals of the All-Union Exhibition of Economic Achievements of the USSR.

Equally important are also scientific-reference complex regional atlases of the Republics, Territories and regions of the USSR, which are a summary of knowledge about the nature, population, economy, culture, administrative set-up and history of these territories. The stock includes complex atlases of many constituent republics and of a number of autonomous republics, territories and regions of the USSR.

This group also includes complex atlases of continents, big regions of the world and separate foreign countries, which, apart from nature maps, contain many original socio-economic maps. These are the Atlas of Africa, the Atlas of Latin America, the Atlas of the USA, and others whose appearance was dictated by the development of international relations.

Numerous and varied is the group of sectoral maps and atlases whose preparation was dictated by the necessity of studying the natural and socio-economic conditions of the country to ensure rational distribution of productive forces and cultural construction. These include cartographic works devoted to the relief of the earth surface, geological structure, climate, hydrology, soils, vegetation, economy as a whole and its sectors, distribution and composition of the population, health service, education and other aspects of natural and socio-economic conditions. The cartographic stock has a considerable number of these editions. For instance, geological maps issued during the last decade are over 300 in number, economic maps - 150 and climatic and soil maps - over 50.

Equally numerous are reference political-administrative maps which are an essential source of information for the mass reader. These include political maps of the world and political-administrative maps of the USSR and its parts. A large share of the political-administrative maps are those of republics, and administrative maps of territories and regions of the USSR.

The stock also contains a considerable collection of reference general geographic maps of foreign countries. Such maps are available to almost all countries.

One of the largest groups in the stock is that of educational maps and atlases.

The group included all standard educational geographic and historical atlases both in Russian and other languages of the peoples of the USSR, as well as varied educational maps: physical, political, political-administrative and administrative maps, maps of natural phenomena devoted to climate, natural resources, soils, vegetation, the animal kingdom, etc., and socio-economic maps devoted to economy, means of communication, electrification, industry, ethnography, history, etc.

A special series of Soviet atlases also intended for the educational process is the educational-regional atlases whose publication was necessitated by the study of the new course "My Region" in senior forms at school.

Complex content of educational-regional atlases which in popular form gives an idea about the nature, economy, culture, population, administrative structure and the history of the territory is also useful for a broad range of readers studying their region.

Of the other groups of cartographic editions mention should be made of the following: maps and atlases devoted to the life and work of the classics of Marxism-Leninism; maps about the Communist Party of the Soviet Union; maps describing the events of the Great Patriotic War of the Soviet Union; maps showing the achievements and prospects of the development of the national economy and culture of the Soviet Union.

Apart from scientific-reference and educational maps, the cartographic stock includes numerous tourist maps and schemes of different regions of the Soviet Union and tourist-excursion itineraries - railway, water, automobile - unmounted and combined.

In recent years the cartographic stock has been enriched by relief maps which represent a three-dimensional projection of the terrain.

The rich collection of Soviet maps and atlases stored at the cartographic department of the Lenin State Library shows the achievements gained by the Soviet Union in the study of nature and socio-economic and historical phenomena, and testifies to the wide use of the cartographic method in research.

R u s s i a n m a p s a n d a t l a s e s

The collection of Russian maps and atlases of the Library is second in scope only to that of the Saltykov-Shchedrin State Public Library.

Of great interest are the atlases of the 18th century which are historical monuments of Russian cartography.

Russian atlases in the stock are represented by the following groups: general geographic reference atlases, sea atlases, atlases of travels, educational geographic and historical atlases, as well as thematic and complex atlases.

Of the general geographic reference atlases of the 18th century mention should be made of the 1st atlas of Russia prepared on the basis of instrument photography and covering the whole territory of the Russian state: "Atlas Rossiiskoi, sostoyashchei iz devyatnadtsati spetsialnykh kart, predstavljayushchikh Vserosiiskuyu imperiyu s pogranichnymi zemlyami, sochinennoi ... staraniyem i trudami Imperatorskoi Akademii nauk" (1745) [Atlas of Russia composed of nineteen special maps of the Russian Empire and borderlands, produced ... by the Imperial Academy of Sciences], and of the 2nd reference atlas "Rossiisky Atlas iz 44 kart sostoyashchii i na sorok na dva namestnichestva imperiyu razdelyayushchii" (1792) [Atlas of Russia consisting of 44 maps and dividing the empire into 42 vicegerencies] reflecting the administrative structure of Russia and compiled on the basis of the General land-surveying.

Of the first regional reference atlases the stock included the atlas of the Orenburg territory "Landkarty ili chertezhi geograficheskiye ..." (1755) [Land maps or geographic drawings ...], "Atlas Kaluzhskogo namestnichestva" (1782) [Atlas of Kaluga vicegerency], and others.

A large group of general geographic atlases is made up of world atlases, atlases of the Russian empire and regional atlases of the provinces of the Russia of the 19th & early 20th centuries. These include A. Ilyin's "Podrobnyi atlas vsekh chastei sveta" (1866) [Detailed atlas of all the continents], which was regarded at one time one of the best reference world atlases; "Bolshoi vsemirnyi atlas Marksa" (1905, 1906, 1916) [Marx Great World Atlas] edited by Prof. E.Yu. Petri and Prof. Yu.M. Shokalsky, which is the main aid for studying geography of pre-revolutionary Russia; "Geograficheskii atlas Rossiiskoi imperii ... Soch ... Pyadyshevym ..." (1823) [Geographic atlas of the Russian empire ... Compiled by ... Tyadyshev ...] based on the materials of topographical surveys and general land-surveying; "Topograficheskii mezhevoi atlas Tverskoi gubernii" (1863) [Topographic land-surveying atlas of the gubernia of Tver] composed as a result of triangulation work and instrument surveys of roads.

The earliest geographic atlases of Russia include navigation atlases which are represented in the Library's stock by such editions as "Kniga razmernaya gradusnykh kart Ost Zee ili Varyazhskogo morya" (1714) [Sized book of degree maps of Ost See or the Varangian Sea ...] which is the first marine atlas of Russia, and "Atlas vsego Baltiiskogo morya s Finskim i Botnicheskim zalivami ..." (1791) [Atlas of the entire Baltic Sea with the Gulf of Finland and the Gulf of Bothnia ...]. The Library's marine atlases of the 19th century include "Atlas Belogo morya s prilezhashchimi zalivami ..." (1825-1827) [Atlas of the White Sea with bordering gulfs ...], "Atlas Kaspiiskogo morya ..." (1860) [Atlas of the Caspian Sea...] and other marine atlases based on marine inventories.

Of considerable interest for the history of domestic geography and cartography are the atlases devoted to magnificent voyages of Russian seamen and scientists, which had led to most important geographic discoveries in the Arctic and Pacific oceans. These atlases include "Sobraniye kart i risunkov prinadlezhashchikh k puteshetviyu ... Yu. Lisyanskogo na korable Neva" (1812) [The collection of maps and drawings concerning the voyage ... of Captain Yu. Lisyansky on board the "Neva"], "Atlas k puteshestviyu vokrug sveta kapitana Krusenshterna" (1813) [Atlas to the round-the-world voyage of Captain Krusenstern], "Atlas k puteshestviyu Kotsebu ... v Yuzhnoye morye i Beringov proliv" (1821-1823) [Atlas to the voyage of Kotzebue ... to the South Sea and the Bering Straits], "Atlas k puteshestviyu vokrug sveta shlyupa Senyavina ..." (1835) [Atlas to the round-the-world voyage of Senyavin's sloop], and others.

The collection of the Library's Russian atlases includes a large number of educational geographic and historical atlases which were issued as aids for the geography of parts of the world and of Russia, as guides in ancient, Middle Ages and modern history, as well as in the history of the Russian state. These atlases were intended to meet various user needs: those of secondary educational establishments, provincial high schools, district colleges, as well as those of juveniles and beginners.

The first educational atlases appeared in Russia due to the expansion of the network of schools providing general and more secular education. These atlases include "Detskii atlas ili novyi i dokazatelnyi sposob k ucheniyu geographii ..." (1768) [The children's atlas or a new and demonstrative way of studying geography ...], the first educational Russian atlas for people's colleges called "Obshchiye i chastniye karty prinadlezhashchiye k kratkomu zemleopisaniyu Rossiiskogo gosudarstva" (1787) [The general and particular maps giving a brief description of the land of the Russian state], and "Atlas sochinennyi k polze i upotrebleniyu yunoshestva i vsekh chitatelei vedomostei i istoricheskikh knig" (1737) [Atlas to be used by juveniles and all readers of gazettes and historical books] intended for students of academic high schools.

Rapid development of many fields of science in the 19th century connected with great changes in the socio-economic and cultural life necessitated the creation of special thematic atlases in Russia. The cartographic department of the Library holds a collection of different thematic atlases of Russia, such as "Gidrograficheskii atlas Rossiiskoi imperii" (1832) [The Hydrographic atlas of the Russian empire ...] containing maps and drawings of the existing and projected water systems of the European part of Russia; "Klimatologiskei atlas Rossiiskoi imperii isdannyi Glavnoi fizicheskoi observatoriyei v pamyat pyadesyatiletnei yeyo deyatelnosti 1849-1899" (1900) [The climatic atlas of the Russian empire issued by the Main physical observatory to mark the 50th anniversary of its activities] which is a fundamental cartographic work about meteorological characteristics of the territory of Russia; the first economic atlas of the country "Khozjaistvenno-statisticheskii atlas Yevropeiskoi Rossii"[4)] [Economic and statistical atlas of European Russia] devoted to agriculture; and the "Atlas aziatskoi Rossii" (1914) [Atlas of Asiatic Russia] which is a thorough cartographic work describing a complex of elements of historical, physical and economic geography of Asiatic Russia and belonging to the best achievements of pre-revolutionary Russian cartography.

The Library's cartographic collection also holds numerous Russian maps characterizing many-sided development of home cartography from the 18th to the early 20th centuries.

Of the maps of the 18th century mention should be made of general geographic maps of the whole of Russia, its border lands and international territories, separate regions and provinces, as well as road and navigation charts and educational maps. These maps include "Generalnaya karta Rossiiskoi imperii s razdeleniyem na novouchrezhdennyye gubernii i uyezdy" (1785) [The general map of the Russian empire showing the newly established provinces and districts], "Pochtovaya karta Rossiiskoi imperii" (1760) [The postal map of the Russian empire], "Karta Ingermanlandii i Karelii" (1740) [The map of Ingermanlandiya and Karelia], "Karta Severnoi chasti Novgorodskoi gubernii s yeye provintsiyami i uyezdami" (1772) [The map of the Northern part of the Novgorod region with its provinces and districts].

The collection of Russian maps of the 19th & early 20th centuries is more varied and is represented by general geographic and various thematic maps devoted to the climate, soils, geology, mineral resources, means of communication and transport, economy, industry, agriculture, ethnography, etc. of different parts of Russia and of the whole world. These maps include "Podrobnaya karta Rossiiskoi imperii i blizlezhashchikh zagranichnykh vladenii" (1801-1804) [The detailed map of the Russian empire and neighbouring lands] which is an outstanding example of many similar editions; "Karta promyshlennosti Yevropeiskoi Rossii" (1842) [Economy of the European part of Russia] which is a valuable source for studying geography of pre-reformed Russia; the "Gipsometricheskaya karta Yevropeiskoi Rossii" (1889) (Hypsometric map of European Russia) which was the first to have given a correct and scientifically grounded idea of the relief of the European part of Russia.

Educational geographical and historical maps which were issued, like atlases, for teaching geography and history figure prominently among the Russian maps of the 19th and early 20th centuries.

The collection of Russian maps and atlases issued before 1917 is of special value, as they were rarely reprinted in pre-revolutionary Russia, and the editions were small.

Foreign maps and atlases

The collection of foreign cartographic works of the Library is the largest in the Soviet Union both in composition and scope.

Of the greatest value is the collection of atlases. These are large and small world atlases, atlases of big regions of the world, complex national atlases of foreign countries and complex atlases of the regions of the world, complex atlases of big cities, various thematic atlases (geological, climatic, economic, road, linguistic, historical, etc.) and tourist atlases.

Of the large world atlases the collection has world scientific-reference atlases of the classical type in the form of collections of general geographic maps with extensive indexes of geographical names, just like the English "The Times Atlas of the World", The Italian "Atlante internationale del Touring Club Italiano" and the German "Haack Weltatlas".

Apart from scientific-reference atlases, the cartographic stock also includes great world atlases which are original, well illustrated encyclopaedias combining the collection of geographical maps, brief annotations and tables, as well as spectacular coloured photographs. This can be exemplified by the atlas "The Earth and Man.A Rand McNally world atlas".

The collection also includes small, so-called pocket world reference atlases containing, as a rule, reference political and physical maps and certain reference-statistical data on the geography and economy of foreign countries.

Of great value are national complex atlases of foreign countries. The national atlases generalize modern knowledge about the geography of the country and reflect the general level of the development of national science. The cartographic stock includes atlases of many countries: Australia, Austria, the Bulgarian People's Republic, Canada, Cuba, the Czechoslovak Socialist Republic, Great Britain, the Hungarian People's Republic, Japan, Netherlands, Peru, Spain, Sweden, Switzerland, the USA, Zambia and others.

Apart from national (state) atlases, another group of cartographic works includes regional atlases describing parts of the country - provinces, regions, etc. Among these works are the "Atlas regionalny wojewodstwa Olsztynskiego","Atlas slivenski okrug", "Az Alfold foldtani atlasza terkepsorozat", "Atlas de Normandie", "Selected maps from the Planning atlas of the Demodor Valley Region".

The cartographic department holds a substantial collection of complex atlases of capitals and big cities of the world. These atlases provide graphic spatial information about numerous social and economic factors affecting the development of a city. Examples of such atlases are the Atlas of Greater London, the Atlas of the city of Linz, the Atlas of New York and others.

Equally varied in composition is the collection of foreign maps ranging from topographic and general geographic to most varied thematic maps covering many aspects of nature and society. These are scientific-reference, educational and tourist maps. Foreign relief maps make up a small collection in the Library's stock.

Apart from maps and atlases, the foreign section of the cartographic stock holds a collections of plans of foreign cities, including separate plans and collections of plans in the form of atlases of general geographic and special content.

International editions make up a special group in the cartographicstock, They include the World Map, 1 : 2 500 000 prepared by cartographic services of seven socialist countries-Bulgaria, Czechoslovakia, The German Democratic Republic, Hungary, Poland, Rumania and the Soviet Union; a number of geological maps - the International tectonic map of Europe, 1 : 2 500 000 and the International map of coal-bearing deposits of Europe, 1 : 2 500 000 prepared by geological departments of many countries, including the Soviet Union; and the Bioclimatic map of the Mediterranean, a complex atlas of West Africa, the Atlas of the biological resources of the world and other international editions prepared by UNESCO.

The cartographic stock of the Lenin Library which is varied in content and purpose is easily accessible due to the fact that there is plenty of information about it.

The broadest information about cartographic works is provided through a system of card catalogs and card files which covers the entire cartographic stock of the Library.

Information about new Soviet maps and atlases is provided through printed cards of centralized cataloging. The newly received foreign cartographic editions are covered in the annual "Catalog of foreign maps and atlases received by the Lenin Library in the year ..." Certain parts of the cartographic stock are covered in a series of thematic printed catalogs, such as "Population maps", "Utilization of soils", "Toponymic maps", "Foreign cities", "Maps of a service area", and other printed catalogs of maps and atlases.

Building-up the cartographic stock

Building-up the Library's cartographic stock aims at shaping a stock which promotes the development of cartography and many other sciences using cartographic editions as sources for cognition and study of spatial distribution of various phenomena in nature and society.

The building-up of each component of the cartographic stock - Soviet editions, domestic editions issued before 1917, and foreign editions - has certain peculiarities.

The collections of Soviet maps and atlases has been shaped largely due to the acquisitions of the obligatory (control) copy. The acquisition by the Library of three copies of all kinds and types of maps and atlases has led to the accumulation of considerable files of rarely requested editions, which affects the preservation of the stock and its use. Therefore, the main problem in supplying the Library with Soviet maps and atlases is to determine how many copies of cartographic works should be permanently stored.

This complex task can be solved only by a thorough and all-round study of reader demands for maps. Reader needs are a mirror, as it were, reflecting modern information needs of many institutions of the country and various reader categories.

The first stage of the study of the user needs made by the cartography department has shown that, while on the whole readers make wide use of cartographic works, demand for certain kinds and types of maps and atlases is uneven. Scientific-reference maps and atlases, as well as tourist cartographic works are in the greatest demand. Educational maps kept at the Library are in much lesser demand. In working out the basic document regulating the building-up of the stock - "Profiles for the enrichment of specialized cartographic stock with domestic printed matter" - these peculiarities of the user needs were taken into account, and the number of obligatory copies to be sent to the Library was determined for various groups of cartographic editions.

According to the new acquisition profile, the Library receives three obligatory copies of all scientific-reference maps and atlases as well as of tourist maps and schemes. Educational maps and atlases for secondary schools are stocked in two copies, and for elementary schools in one copy.

Russian maps and atlases. The collection of domestic maps and atlases gathered at the Library was based more on retrospective than current acquisition, as the obligatory copy law was not fulfilled properly enough.

The main task in supplying the cartographic stock with domestic maps and atlases issued before 1917 is to fill the existing gaps.

This task can be solved by exchanging maps with large libraries of the Soviet Union and purchasing the most valuable editions, as well as by enriching the stock with coloured photocopies, microfiches and microfilms of maps and atlases.

Foreign maps and atlases. The collection of foreign maps and atlases was shaped largely by receiving donations from institutions and private persons, as well as through international book-exchange, purchase of maps and atlases at home and abroad.

The problem of building up the Library's stock of foreign maps and atlases is very complex. It lies, above all, in an ever increasing volume of cartographic works published in the world, which is accounted for by an ever growing importance of the map as a very rich source of varied information. Therefore, one of the main tasks in building up the stock is to identify the correlation between the volume and content of the maps and atlases published in the world and to determine the optimal composition of acquired editions.

Current acquisition of foreign maps and atlases for the cartographic stock can be effective by taking into account the subject matter and prospects of the development of scientific research at the country's institutions which make wide use of the cartographic method of research.

Apart from reader demand, another factor - coordination of acquisition with the country's large libraries - is also very important in optimizing the acquisition of foreign maps and atlases for the Library.

Coordination makes it possible to reasonably distribute cartographic editions among large libraries in order to ensure that a sufficiently complete repertoire of all valuable maps and atlases issued in the world is obtained.

Coordination in building up the cartographic stock should be effected in stages - by kind of edition, by type of maps and atlases, by their subject matter and other characteristics of a cartographic work.

The approach to coordination must be different depending on the subject matter of cartographic editions and the reader demand for them in each library. For instance, big reference geographical and complex national and regional atlases which are in the greatest demand at libraries due to their enormous information capacity should be acquired by all the big libraries of the country. As for the acquisition of cartographic editions, devoted, for instance, to geology or linguistics, coordination is desirable. It is not necessary for all libraries to acquire multi-sheet topographic maps covering all the regions of the world.

Valuable cartographic editions are acquired by one library in the original, whereas other libraries can get copies (photocopies, microfilm, microfiches, etc.) and facsimile editions.

The availability of cumulative catalogs of maps and atlases is a most important precondition for successfull coordination of acquisition.

Of great importance is the establishment of map-exchange with national libraries of the world, with cartographic information centres and with big special and universal libraries of foreign countries.

NOTES

1) *The account of the Moscow Public Rumyantsev Museum for 1867-1869, p. 237.*

2) *The Book of new accessions to the Library of the State Rumyantsev Museum for 1925.*

3) *Proceedings of VTsIK, 1919, March 19.*

4) *The Lenin Library has the 2nd edition of this atlas issued in 1852.*

DEAR MR. NORDENSKIÖLD, YOUR OFFER IS ACCEPTED!

By Ann-Mari Mickwitz

Everyone interested in the history of cartography is acquainted with the *Facsimile Atlas to the early history of cartography* by Adolf Erik Nordenskiöld, published as early as 1889, and still a notable work in its field. Nordenskiöld is well-known for his cartographic studies, which resulted in the publishing of the Atlas and its complement the *Periplus*; and he also gained lasting fame as an explorer of the Northern Polar Sea, and above all as the discoverer of the North-East Passage.

Adolf Erik Nordenskiöld was born in Helsinki, Finland, on the 18th November 1832. His father, Nils Nordenskiöld, was a mineralogist, and Adolf, as a young boy, followed him on mineralogical expeditions to different parts of Finland. Later he studied mineralogy and geology at the University of Helsinki, at that time the Imperial Alexander University (for Finland was then a part of the Russian Empire). Appointed lecturer in mineralogy at this University, Nordenskiöld in 1857 incurred the disapproval of the Russian authorities and had to leave his native country. He went to Sweden where shortly after he was appointed professor at the Royal Swedish Academy of Sciences in Stockholm. It was not easy for the young Nordenskiöld to accept exile from his homeland, but in the event this misfortune turned out to be of utmost importance for his future career. Swedish scientists were at that time especially interested in the study of Polar regions, and Nordenskiöld was able to participate in their expeditions to Spitzbergen and soon became their leader. This led him to study not only the geological, meteorological and marine conditions of these parts of the world, but also the history of navigation in Polar waters. The Spitzbergen area was only one of various destinations of Swedish expeditions. Nordenskiöld planned and led two major explorations to the mouth of the River Jenisey to study the possibility of sailing through the Kara Sea without being stuck in the ice. His studies of old journals of travel and descriptions of navigations in those waters had convinced him that this route was navigable, and these expeditions proved him right.

The Russians also were eager to gain experience from these navigations. Nordenskiöld therefore received financial support from both Sweden and Russia for his expeditions eastward. His interest in the history of Polar fishing, whaling and wintering led him to collect early literature on northern regions. Before he undertook his great expedition for the discovery of the North-East Passage he had brought together a marvellous collection of books, old maps and charts pertaining to the Polar regions. Later on, after his return from the circumnavigation of Asia and Europe, he devoted himself to the study of the history of cartography, collecting more and more old geographical atlases and books on cartography. Eventually, he was able to say that his collection in this field, dating from the 15th to the 17th centuries, was one of the most complete outside the great national libraries such as the Bibliothèque Nationale in Paris, the Library of the British Museum in London and the Library of Congress in Washington, D.C.

Such was the background to Nordenskiöld's collection, now belonging to the University of Helsinki, and one of its library's most precious possessions.

Many of the visitors who come to admire the treasures of the Nordenskiöld collection or to pursue the study of cartography and the history of navigation, ask how Nordenskiöld managed to collect so many valuable and rare items. In the course of his many and various travels he always visited antiquarian bookshops. Learning that he was a good client, booksellers began regularly to send him their printed catalogues and wrote to him about new acquisitions which they thought might be of interest.

Many of these catalogues now form part of his collection, preserved together with notes and invoices. Although incomplete, these files still serve as an invaluable source of information about the origins of the collection. In the Library of the Kungliga Svenska Vetenskaps Akademien (the Royal Swedish Academy of Science) letters of antiquarian booksellers are to be found, forming a complementary collection available for this study.[1] The catalogues of the booksellers not only give prices and short notes about the books offered for sale, but also contain important information about these works, with references to relevant literature. Many such references, for instance, can be found to Nordenskiöld's Facsimile Atlas. A catalogue of Jacques Rosenthal in Munich, 1895, refers to this work no less than six times on one page, listing eight mappemondes. Nordenskiöld himself said that over thirty years he had been trying to build up a collection as complete as possible of geographical and cartographical literature from the 15th to the 16th century, that is, until the publication of Ortelius's *Theatrum Orbis Terrarum* in 1570. He extended his collection later to include the 17th century. The collection, he claimed, formed the treasury on which the *Facsimile Atlas* and the *Periplus* were based. He was able to collect all these rare and valuable items through the collaboration of antiquarian dealers on the continent; not only the large, expensive firms, such as Frederik Muller, Rosenthal and Quaritch, but also less important ones in Germany, Italy and elsewhere.

Among the papers in the Nordenskiöld collection in Helsinki are notes and invoices from 88 different antiquarian booksellers in thirteen European countries. Of these 26 are German, 21 Italian, 12 Swedish, 11 French (of which 10 are Parisian), 9 English, with the remaining eight spread over other European countries. In the library of the Academy of Science in Stockholm there are letters from another 16 booksellers, whose correspondence is also preserved in Helsinki.

It is beyond the scope of this paper to trace all the transactions between Nordenskiöld and these booksellers. Instead, I have focused on the most interesting items and have made it my first and main task to find out where and when Nordenskiöld acquired his many editions of Ptolemy's Geography. In these investigations there are problems which cannot be solved. For example, T.O. Weigel of Leipzig on December 31, 1881, sent a bill with the note: Ptolemäus Geographia M 14:-, yet a catalogue from his shop cannot be found for this period. The same problem arises when Georg Lissa of Berlin writes on his invoice: Ptolemaeus Chatalog 5.11 M 20, and this catalogue is missing among the Lissa catalogues in the collection.

The Swedish antiquarian bookseller Klemming apparently had books on commission from continental antiquarian booksellers. Between 1877 and 1882 he charged Nordenskiöld for a Ptolemaeus on behalf of Fidelio Butsch & Sohn No 135, Skr, 50:-, and "Från Harrasowitz (nr) 116 Ptolemaei Geographiae libri 8 Kr 45". Without mentioning where the atlas comes from he merely writes "Alexandrini Cosmographia 18" and later Tolomeus, Nr 76, Lire 7:-, which of course indicates an Italian edition, and finally, 1 Ptolemeus Atlas Kr 3:-, which seems a very small amount. The Danish bookseller H.H. Lynge of Copenhagen once noted a Ptolomaeus on his bill and his price also was Kr 3:-, but this time it must be Danish crowns. With reference to the firm of Fidelio Butsch & Sohn it may be significant that later on, in 1883-84, Nordenskiöld had direct connections with a bookshop named A.F. Butsch in Augsburg.

To begin with the first edition of Ptolemy, that of 1475 without maps, Nordenskiöld has written his signature in the edition with the note on a free endpaper at the beginning of the book, "Köpt i Paris Nov 1881" (purchased in Paris in November 1881). In Leo S. Olschki's catalogue XLIII, Verona, a somewhat incomplete copy of this edition is found at the price of 1,500 frs and is said to be "albo corvo rarier" (rarer than a white raven). The copy Nordenskiöld bought in Paris is bound in a new leather binding with gilt ornaments and inside dentelles and with a large coat-of-arms on both front and back covers. The text on the coat-of-arms reads: "J. Gomez de la Cortina et amicorum Fallitus hora legendo". A smaller copy of the same arms, in red and gold, is pasted on the inner side of the front cover.

The first Ptolemy edition with maps is dated 1462 in the colophon. Recent research has shown that this is a mistake and that the year should be 1477. On August 16, 1888, Olschki in Verona wrote a letter to Nordenskiöld saying that he could offer a beautiful copy of this edition at a price of Fr. 2,250. On August 17 (sic) he expressed his regrets that Nordenskiöld thought the price too high. He said that in Italy and thereabouts this edition was estimated the rarest of all the Ptolemys and indeed he himself had never seen any copy of it in the market nor had he seen it announced in any catalogue.

For him, as a bookdealer, this fact proved its rarity. If Nordenskiöld would let him know what his highest price would be and if the difference was not too much, he would do his best to get the copy for Nordenskiöld. On September 29, 1888, Olschki expressed his regrets that Nordenskiöld had not yet been able to make up his mind about the Ptolemy of 1462, saying that the German bookseller who promised Nordenskiöld a copy at the price of 500,00 (five hundred Frs), would never be able to keep his word. Olschki would be very grateful if he could sell him some dozen copies at that price! He was sure that this edition was of less interest than that of 1478, but it was superior to the later edition. He was also quite certain that an American library collecting Ptolemy editions would pay Frs 2.500 for it, at which price the copy would be found in his next catalogue (nr 94). Nordenskiöld should be aware of the fact that Olschki's offer was already reduced and that the price could not be made lower, as he himself had paid 2,150 Frs for the copy. And on October 29 he announced that the 1462 Ptolemy was sold at the price of 2,500 Frs without any reduction. Now he would do his very best to find another copy and would be very happy indeed to be able to offer it at a lower price. On October 28 Olschki was still speaking of the same edition and wrote that he had received many inquiries about it. He stressed once more that Nordenskiöld must know that the price was not too high. On a carte postale the selling of the famous copy at 2,500 Frs was mentioned once again on October 29. All this gives the impression that Olschki felt rather guilty about not having sold the Atlas to Nordenskiöld. Then on November 3, he was able at last to offer Nordenskiöld a cheaper copy, although not a complete one, as eight maps were missing. The text was complete and with wide margins. One side of the cover was damaged, but the copy could be said to be good. The volume was precious even if some maps were missing, and Olschki thought that he did Nordenskiöld a great service, when he hurried to offer him this copy. He asked Nordenskiöld to reply very soon, as this rarity could be sold very easily even in fragments. The price would be Frs 550, and he thought it so moderate that he hoped to receive an order by telegram, in which case the words "Olschki-Verona-Acceptée" would be enough. In a PS he again emphasized: the copy contained 18 maps and only eight were missing. On November 16 Olschki wrote that he had received Nordenskiöld's telegram and that he would hurry to send the 1462 Ptolemy by train.

Now there followed a letter from Nordenskiöld to Olschki, in which he acknowledged the receipt of the volume, but maintained that the last leaf was missing and that this leaf was of the greatest importance as it contained the imprint with the names Manfredus, Petrus, Borenius etc. In these circumstances he would not pay more than 450 francs. In his answer Olschki said that on comparing the incomplete copy with the complete one (sold at 2,500 Frs) he had found that the texts were absolutely identical, and ended with the words "Anno MCCCCLXII, Mense Iunii, XXIII, Bononiae". Olschki was very unhappy about the matter and wondered if the imprint would be found before the descriptions. He could not in any event sell the volume for 450 Frs, as he had himself paid 500 Frs for it.

There is more correspondence about this unfortunate business. Nordenskiöld sent Olschki 450 Frs, but Olschki would not leave it at that, as he did not like to lose 50 Frs in selling the book. On January 12, 1889, Olschki at last thanked Nordenskiöld for sending him Frs 70,00 to pay the credit of his account. One might presume that the 1462 Ptolemy deal was now settled, but a final episode can still be reported. On May 3, 1901, Olschki wrote a letter saying that he had once sold Nordenskiöld an incomplete copy of the Ptolemy Atlas of 1462. He now wondered if Nordenskiöld had acquired a better copy elsewhere. If so, Olschki had a client who would be willing to pay 1,500 Frs or even more for the incomplete copy and would Nordenskiöld part with it. That is the end of the story. The Nordenskiöld collection still contains the volume which was the subject of this correspondence.

Along with all these arguments about the 1462 (1477) edition of Ptolemy, there was correspondence about the 1478 Rome edition. Olschki offered this edition for the first time on May 5, 1888, his price being 1,500 Frs. The copy in question was bound in a very fine morocco binding, one leaf in facsimile. Olschki said that the maps were the oldest existing engraved in this manner and also the most beautiful ever made for a Ptolemy edition. Later Olschki wrote that he would reduce the price for the Ptolemy by 100 Frs, hoping that his correspondent would accept a price of 1,400 Frs, which was indeed a very low price. His correspondent did in fact accept the price, and ordered the copy on July 12, 1888.

On the 18th Olschki said that he would let Nordenskiöld know this. It seems that Nordenskiöld did not accept the offer, for soon after Olschki offered another copy of the same Ptolemy edition in a very special binding, which alone was worth 500 Frs, the price of the whole atlas being 1,500 Frs. As this was a very desirable item, Olschki requested a reply by telegram. On August 16 he then confirmed Nordenskiöld's order and said that he would despatch immediately the splendid copy, very well packed. On September 8 he acknowledged the receipt of the 1,500 Frs. Nordenskiöld has left in the atlas a leaf from Albert Cohn's Katalog 202 Berlin 1892. This catalogue contains the following remark: "182 Ptolemaeus (1478) 2.500, - Livre excessivement rare. Première édition de Ptolémée avec des cartes. Seul volume portant le nom de l'imprimateur Arnold Bucknick." There is an owner's mark pasted on the inner side of the front cover with the name "Henry Stevens of Vermont FSA 4 Trafalgar Square London 1882". And on the tail of the inner side of the front cover there is printed in gold: "Bound by W. Pratt for H. Stevens & Son".

About 1480-82 an edition of the Ptolemy with maps was published in Florence with a metrical paraphrase of the text by Berlinghieri. Nordenskiöld bought this edition from Bernard Quaritch in London for £12.12s. on August 27, 1886 without any argument. The copy belongs to the first issue, which is, according to Stevens (p.38)[2], "Excessively rare". In this first issue the recto of the first leaf and the entire last leaf are blank. Nordenskiöld has pasted in a cutting from an English catalogue with the price 6 300, and has written on it "Detta Exemplar (this copy) £12-12". A card of August 9, 1886 from Silvio Bocca in Rome is fastened on the inner side of the back cover. On the card Bocca offered a "magnifique exemplaire du Berlinghieri avec les planches, mais il en vent 2 000 francs". Nordenskiöld's note on the card is: "Trop cher pour moi, Mr. Quaritch m'offre une copie ...". The invoice from Quaritch of August 27, 1886 states: "Berlinghieri Geographia large folio wanting Map of Palestine bound in oaken board 12 12 per Wilson steamer from Hull. The book is much too large for the mail". And Nordenskiöld's note reads: "Btd 23 Sept. 1886 gm. Skandinav. bolaget". (Paid 23 Sept. 1886). Quaritch acknowledged the receipt on August 29.

The Ulm edition of 1482 cannot be traced to any bookseller. Olschki once offered it at a price of 1,000 Frs, but there is no invoice preserved and the volume contains no marks of provenance. On the other hand it is clear that Nordenskiöld bought his copy of the Ulm edition of 1486 from Frederik Muller in Amsterdam.[3)] The edition is found in Muller's catalogue of 1877 as nr 2617; the sum of 45 - 100 = Florin 41,59 - was sent on September 3, 1877, and on September 27 Muller acknowledged its receipt.

For the Rome edition of 1490 there is only one offer of a copy from Olschki in Venice, this at the price of 650 Frs. The letterhead has now a printed text: "date de timbre postal" and the printed name of Verona has been corrected by hand into Venice. Olschki moved from Verona to Venice sometime in August 1890 and stayed there until 1897 or 1898, when he moved his business to Florence. The offer must consequently have been made between 1890 and 1897-98.

Nordenskiöld bought from Olschki three other less expensive editions of Ptolemy, i.e. the edition of 1520 printed in Strasbourg, for 50 Frs., the Italian edition of 1564 at 36 Frs. and the 1597 edition printed in Cologne and containing the Portolano, printed in Arnhemii, Apud Joannem Jansonium Bibliopolam Anno MDXCVII. Olschki mentions this edition as printed in Arnheim, which seems to be a mistake, as no such edition is known. The price of this edition was 45 Frs. The Nordenskiöld collection contains another copy of this edition, without the Portolano.

There was no discussion about the price for these three editions; Olschki's offers were immediately accepted. One offer from Olschki on December 30, 1893, however, was not accepted, and this is all the more remarkable as the item in question was the Strasbourg edition of 1522, the only one printed before 1600 missing from the collection. Olschki did not mention the price and nothing more is on record in the papers now available.

In addition to the Ulm edition of 1486 already mentioned, Nordenskiöld acquired during the years 1884-89 the following editions from Muller: 1514, 1542, 1545, 1552, 1561, 1564, 1578, 1584, 1597/98, 1617, 1698 and 1828, the prices varying between 11 and 45 florins. The highest price, 45 florins, was paid for two editions, that of 1552 printed in Basel and that of 1578 printed in Cologne, the first Ptolemy edition with maps by Mercator.

Nordenskiöld was also a client of the Roman bookseller Silvio Bocca. From him Nordenskiöld bought the Ptolemy editions of 1511 and 1513, printed in Venice and Strasbourg respectively. On Bocca's letter of December 25, 1892, Nordenskiöld has written "Betalade den af 1511 och erbjudit mig att taga den af 1513 mot 4 ex af Marco Polo", (paid for that of 1511 and offered to take the one of 1513 in exchange for 4 copies of Marco Polo). Bocca's price was 150 francs for the 1511 edition and 250 francs for that of 1513. Nordenskiöld had in 1882 published a facsimile edition of a Marco Polo MS, belonging to the Royal Library in Stockholm, and he offered four copies of this work in exchange for the Ptolemy. According to Bocca's receipt, he accepted two Marco Polos and 130 francs.

Another two Ptolemy editions in the collection can be traced back to their sellers, as follows. Nordenskiöld bought the Rome edition of 1508 from Ludwig Rosenthal in Munich, one of the two Rosenthal booksellers in that town, both of whom had dealings with Nordenskiöld. The 1608 edition came from Dufossé in Paris, from whom also many of the separate maps in the collection were bought.

Some other aspects of Nordenskiöld's dealings with Olschki are of interest. In a postscript to a letter of June 7, 1889, Olschki asked if Nordenskiöld had yet been able to find a copy of Crescentio's *Nautica Mediterranea,* with the geographical map. He now had an offer of one at 225 Frs, and asked whether the price was too high. He stated that it was very difficult to find a complete copy, and his efforts to find one had yielded no results to date. Nordenskiöld must have thought that the price was indeed too high, and nothing was heard about the Crescentio until February 20 in the following year (1890), when Olschki took the matter up again with the same remark, that it was extremely difficult to find a copy with the large map. Now one of his clients was in possession of such an item but was asking 200 Frs for it. According to Olschki's description the copy was in good condition and absolutely complete. Olschki said that during the last three years he had been looking in all available catalogues and desiderata lists for a complete copy for Nordenskiöld, but in vain; a fact which for him was testimony of its great rarity. He also remarked that the price had been only 50 Frs some time earlier but had now risen to 200 Frs and it was not easy to find a complete copy even at that price. On April 14, 1890, he took the liberty of sending Nordenskiöld a copy at 160 Frs and was sure that it would satisfy him. The invoice for the book is dated April 14 and the bill was paid on the 22nd of the same month. Nordenskiöld has written his signature and the date, April 1890, on a leaf at the beginning of the book and on the paste-down endpaper: "162 fr, (which includes the postage), Köpt af Olschki i Verona" and "In Fine Portolano della Maggior parte d'luoghi da stantiar navi, et galee in tutto il mare Mediterraneo - In Roma MDCII". The Mediterranean section of the text is lacking in the other copy of the work preserved in the collection.

It is natural to ask: where did Nordenskiöld buy the many other valuable atlases and books in his collection? I shall answer this from selected examples.

Three of the Mercator atlases came from Olschki, i.e. the first edition of 1595, and two atlases for which no edition is mentioned. As usual Olschki had some comments to make in connection with two of these purchases. Remembering the argument about the imprint of the 1462 (1477) Ptolemy he asked Nordenskiöld to look for the date and the printer's name on the Atlas of 1595 at the end of the second part. About the second Atlas he wrote on a card in English: "The edition of the 16. century sold to you was paid much more. (90 fr), because all booksellers are seaking now editions of the 16th century relating to America". The price of the second Atlas was only 60 frs, but it was a large one, weighing 8 kg, and to send it as cheaply as possible Olschki brought it to the Austrian border from Venice and sent it thence by mail to Stockholm. He claimed to have paid 14.80 frs for the mailing but charged Nordenskiöld only 10 frs. Although the edition of this atlas was not specified, it must have been one of 1630, as the collection contains only two Mercator atlases in large folio, and the other one, that from 1613, was acquired in 1886 from Frederik Muller.

For ten of the Ortelius atlases in the collection there are invoices with the name of the booksellers and the price of the atlas, but three of them do not mention which edition they were selling. Muller and Olschki both sold three editions, Bocca two, while Otto Harrassowitz in Leipzig and Max Kantorowicz in Milan each sold one edition. The most expensive edition was that of 1570, sold by Muller for 60 M.

In Nordenskiöld's own opinion the Lafreri atlas *Tavole moderna di geografia de la maggior parte del mondo ...* was the most valuable of all the atlases in his collection. He bought this atlas from Olschki in February 1888 for 425 Frs after having tried in vain to give nine copies of his Marco Polo in exchange for it. The negotiations went on from July 1887 to February 1888, the very last letter mentioning the Lafreri being one of March 20, 1888. Nordenskiöld's own note in the atlas itself reads:"Köpt mars 1888 hos Olschki i Verona för 434 fr" (Bought in March 1888 from Olschki in Verona for 434 frs). The copy contains the Carta marina of Olaus Magnus, printed in Rome by Lafreri in 1572, a map often missing from Lafreri atlases.

There is another Italian atlas factice, which came to be called Lafreri II of the collection. In actual fact it is not a Lafreri atlas; the Italian maps were collected and bound together in Poland about 1570. Nordenskiöld acquired this item from A. Thury in St. Petersburg in November 1894, the price being £10.

The large *Atlas Major* of Joan Blaeu, 11 volumes in magnificent original binding, was sold to Nordenskiöld by Muller. The first volume is from the 2nd edition of 1665, whereas the other ten belong to the 1662 edition. The condition of the atlas is excellent, and the colouring brilliant. The price was 70 Frs when bought in 1877. Another of Blaeu's atlases, the *Zeespiegel,* Amsterdam 1627, at a price of 8 florins, came from Muller's bookshop in November 1889.

Marine atlases sold to Nordenskiöld by French antiquarian booksellers include Ch. Chademant Bellin's *Le petit Atlas Maritime* in five volumes, Eduard Dufossé *Le grand & Nouveau Miroir ou Flambeau de la Mer.* [Theunitz Jacobscz] and a Colom atlas from Maison-neuve & Leclerc.

Among the treasures in Nordenskiöld's collection there are examples of atlases known today in only a very few copies. For example, the collection contains some rare works of Vincenzo Coronelli. Nordenskiöld bought from Olschki in January, 1900, 13 volumes of Coronelli's *Teatro della guerra*, some of whose volumes are now found in only a few other libraries. In addition to this series there is also another volume of *Spagna* bound together with the *Archipelago*, which Nordenskiöld acquired from Basilio Benedetti in Rome, with whom he had transactions during the ten years 1890-1900.

A small book much valued by Nordenskiöld was Giovanni Scandianese's *I Quattro libri della caccia*, in Venezia, 1556. He wrote on a leaf in the book: "Arbetet mycket sällsynt. Innehåller s. 23 en märklig karta äfven omfattande den Nya Världen, i Macrobius stil" (A very rare book. Contains on p.23 an extraordinary map also comprising the New World in the style of Macrobius). In the Facsimile-Atlas, p.37, under the world map of Ymago mundi of Petrus de Aliaco (Pierre d'Ailly), c.1483, Nordenskiöld made almost the same remark in English in a MS note: "A very interesting map of this type is inserted in Giovanni Scandianese, I Quattro libri della caccia. In Venezia 1556 p.23. It forms a hemisphere of 72 m.m. diam. surrounded by 8 wind-heads and embracing the Atlantic, Europe, Africa, a part of Asia and the New World. The whole is crossed from "tropico di Cancero to tropico di Capricorno" by "Zona torrida inhabitabile". It is the only map I can remember where "Zona torrida Inhabitabile" is extended to the New World"(bör reproduceras)" "(it ought to be reproduced)." Olschki had already offered this book to Nordenskiöld in 1894, but there is a note on the letter: "har ej köpt den". In the year 1897 both Jaques Rosenthal and J. Halle in Munich offered the book. Halle sends it "Zur gefl. Ansicht" on November 29 and writes: "Interessantes Jagdbuch, das Seite 23 eine mapmonde enthält mit Nueva Hispania, die in Ihrem Facsimile-Atlas nicht enthalten ist. M. 30". Rosenthal's offer is dated October 1897, the price being 75 M.

A very rare and precious book is the *Isolario* of Bartolomeo degli Sonetti, dating from 1485, comprising sailing directions for the Aegean Sea, written in sonnets. The small maps, numbering 49 in our copy, depict the islands of this area.

Nordenskiöld bought the book in December 1892, from Ulrico Hoepli in Milano and paid 350 frs. for it. Isolarios, as such books about islands were called, were of course very popular in earlier centuries. Nordenskiöld acquired from Olschki the first edition of Bordone's *Isolario* 1528, one of his Porcacchis from Hoepli, and another from Olschki. Marco Boschini's *L'Archipelago*, 1658, he bought from S. Rosén in Venice, 1894, for 30 fr. 50 c., as he records in a note on the inner side of the back cover. Olschki speaks about an edition from 1580 of Ferretti's *Diporti notturni*, which he had offered to Nordenskiöld in 1889. Such a copy is found in the collection. Of the six editions of Lucas Jansz Waghenaer's *Spiegel der Zeevaert*, Nordenskiöld bought the Latin edition of 1586 from Francesco Casella Gomaro in Naples and all the others from Frederik Muller; the first edition of 1584-85 in Dutch, the German from 1589, the French from 1590, a Latin from 1591 and another Dutch edition with the year 160., the last figure missing but given by Muller as 1605. Muller's prices varied between 85 fl. for the first edition to 40 fl. for that of 1590.

Nordenskiöld's biggest deal with Frederik Muller dates from December 1900, when Nordenskiöld wrote to Muller announcing that he intended, in a few years, to move from Stockholm to his home in the countryside. He was not able to take the whole of his large library with him and, if he was not very much mistaken, Muller two or three years earlier had offered to buy the entire collection. At that time Nordenskiöld had not been ready to consider such a transaction but he was now prepared to do so. Nordenskiöld went on to describe the value of the collection and its treasures. He pointed to his contacts with large and small antiquarian bookshops, which had enabled him to acquire all these items and he enumerated many of his most remarkable books. His price for his entire collection would be 300,000 Frs. He asked Muller to answer as to wheather he would be prepared to undertake this transaction, but insisted that he would sell the collection only as an entity and at the price mentioned.

Although Nordenskiöld's offer was made in December 1900, there must have been some earlier negotiations, as there are letters in the collections of the Royal Swedish Academy of Science from Muller, from the Buchhandlung Gustav Fock in

Leipzig and from Otto Harrassowitz. In July-August 1899 all of them refer to information in the press about Nordenskiöld's intentions of selling his library. Fock wrote directly to Nordenskiöld and promised him "eine besonders vorteilhafte Verwertung der Bibliothek". Harrassowitz wrote to the publisher John E. Bergsjö in Stockholm asking him for more information and for assistance in getting in touch with Nordenskiöld. He was very anxious to secure the commission and promised Mr. Bergsjö an appropriate percentage if he should succeed. The letter was forwarded to Nordenskiöld.

In a reply of January 8, 1901 to Nordenskiöld Muller wrote that he agreed in principle to the proposal of December 29, 1900. But he would like to settle some points:

1) That Frederik Muller & Cie would be the only firm to sell the collection at the stated price;
2) That he should have the right to let possible buyers know Nordenskiöld's name, because they would probably guess it anyway;
3) What percentage would Nordenskiöld agree to if the transaction were to be successful? (Here Nordenskiöld has inserted 10.)
4) Would Nordenskiöld have any objections to signing a contract, for example, one with a fixed time-limit?

In a reply to Nordenskiöld's letter of January 15, Muller made some further comments. He needed more time for handling the matter, as his correspondents were located in Australia and America amongst other places, the minimum time being one year. He also needed a printed catalogue of the collection. This could be prepared by two members of his staff, who would come to Stockholm for that purpose. He emphasized that this would not be the first business of the kind handled by his firm, that the catalogue was absolutely essential, and that the sale could not take place within one or two months. Nordenskiöld then extended the term to the end of April 1902 and he himself wrote a paper "Notes sur la bibliothèque de Mr. Nordenskiöld". Muller, however, could find buyers only for parts of the library and nobody was prepared to take over the whole collection. Muller had many replies from Melbourne and Sydney and in a letter of March 23, he enclosed copies of notes from Harvard University and the Lenox, Astor & Tilden foundations in New York, all of which wanted to buy items which they lacked. To this letter he never received a reply. Nordenskiöld died on August 12, 1901, and a letter of condolence from Frederik Muller to Mrs. Nordenskiöld, dated August 16, 1901 is the very last item of the correspondence.

Thus no antiquarian bookseller took over and disposed of the valuable collection of Adolf Erik Nordenskiöld. He himself had sometimes expressed a hope that his library could find its way to his native country and this also was the sincere wish of his family. In the event, the collection was offered to the University at Helsinki and finally became part of its library. Thus Helsinki University Library came to possess one of the finest collections in the world on Renaissance cartography.

NOTES

1) *Brev till A.E. Nordenskiöld. Serie I, and Tillägg till brevkatalogen.*

2) STEVENS, HENRY N.: *Ptolemy's Geography, A brief account of all printed editions down to 1730. 2nd ed. London, 1908.*

3) *There is an essay about Nordenskiöld's relations with Frederik Muller in Amsterdam written in Finnish:* NIVANKA, EINO: *Amsterdamilainen Fr. Mullerin Antikvariaatti Nordenskiöldin kirjakokoelman päähankkijana (The bookseller Fr. Muller & Co. of Amsterdam as chief purveyors of books for the Nordenskiöld Collection). (Miscellanea bibliographica 5, 1947. Helsingin yliopiston julk. 20.)*

4) NORDENSKIÖLD, A.E.: *Om ett aftryck från XV: de seklet af den i metall graverade världskarta, som förvarats i kardinal Stephan Borgias museum i Velletri, (Ymer, 11, 1891, p.83-89).*

VIGLIUS AB AYTTA ZUICHEMUS, SAVANT, BIBLIOTHECAIRE ET COLLECTIONNEUR DE CARTES DU XVIe SIECLE

Par Antoine De Smet

Viglius ab Aytta est né le 19 octobre 1507 sur la terre de Barrhuys, près de Wirdum, au sud de Zwichem en Frise (entre Leeuwarden et Sneek) et mourut à Bruxelles le 18 mai 1577. Il ajouta "Suichemius" à son nom en souvenir du village Zwichum ou son oncle Bernardus Bucho van Aytta, qui l'avait adopté, fut curé.

Grâce à la protection intelligente, aux moyens financiers et aux influences dont disposait cet oncle, le jeune Viglius parcourut brillamment un cycle d'études sur les traces des grands humanistes de son temps. En 1519 il se rendit à l'école de Deventer; en 1520 il étudia à La Haye où son oncle était doyen du chapitre. La même année 1520, il reçut la tonsure à Utrecht, ce qui montre qu'il envisagea une carrière ecclésiastique.

A Leyde, il poursuivit ses études dans les langues classiques sous la direction de Gulielmus Verius. A La Haye, il fut ensuite guidé par Jacobus Volkert de Bergen. Lorsque ce savant fut nommé recteur d'un groupe d'étudiants hollandais à Louvain, Viglius l'y accompagna (1522-1524). L'étudiant Frison était ainsi lancé à fond dans ce brillant mouvement scientifique qui fait la gloire de l'Alma Mater de Louvain dans la première moitié du XVIe siècle. Il y suivit entre autres les cours de latin de Conradus Goclenius (1489 Louvain, 25 janvier 1539) et ceux de grec de Rutgerus Rescius (Maaseik, fin du XVe siècle - Louvain, 2 octobre 1545).

Les études de droit l'attirèrent tout spécialement; il eut comme maîtres Joannes Hagius et Nicolaus de Bruxella.

Que de savants néerlandais et étrangers n'eut-il pas l'occasion de fréquenter! Les possibilités de s'instruire, de connaître les questions qui faisaient l'objet de la curiosité des érudits de l'époque, de se documenter sur les institutions et la situation politique dans les différentes parties du monde connu, étaient à sa disposition.[1]

En 1526, il quitta les Pays-Bas pour se rendre à Dôle; il poursuivit ses études à Avignon chez Alciatus et à Valence où il suivit ce maître pour fuir une épidémie de peste. Il fut promu au grade de docteur en droit en 1529. Après un séjour de deux ans à Bourges, il voyagea par Orléans, Paris, Lyon, Genève jusque Bâle. Il traversa l'Allemagne et l'Autriche pour s'arrêter à Padoue (1531). En cours de route il visita Amorbach et Froben (Bâle), à Fribourg, il rencontra Erasme, à Augsbourg, les Fugger. C'est dire qu'il se fit une brillante série de relations, parmi lesquelles plusieurs pratiquaient les différentes disciplines en honneur à l'époque.

A Padoue, Viglius se fit remarquer par son enseignement des Institutes (1532-1533).

Malgré tous les honneurs et les possibilités de carrière en Italie et ailleurs, il retourna en 1534 aux Pays-Bas, dans sa Frise natale où l'attendait l'héritage de son oncle Bernard Bucho, et où il retrouvait les membres de sa famille et de nombreux amis.

Il passa les années 1535-1537 en Allemagne comme official de l'évêque de Munster et membre de la diète impériale à Spire. Il fut ensuite attaché à l'Université d'Ingolstadt, comme professeur et doyen de la Faculté de droit, finalement comme recteur.

Bien préparé pour les carrières politiques et juridiques, précédé d'une réputation scientifique et juridique considérable, il fut nommé le 12 octobre 1540 conseiller au Conseil Privé des Pays-Bas.[2] Son amitié avec ses compatriotes les frères Janus Secundus, Nicolaus Grudius et Adrianus Marius dont le père, Nicolaus Everts Everardi (Grijpskerke, près de Middelburg 1462 - Malines 1532) avait été président du Grand Conseil de Malines, lui fut certainement utile à cette occasion.[3]

Il en devint président en 1549, fonction qu'il occupa jusqu'en 1569. Entretemps, il accéda également à la présidence du Conseil d'Etat (1554).[4)]

Comme titulaire de ces dignités parmi les plus hautes dans les Pays-Bas, Viglius exerça une influence prépondérante sur l'administration centrale de nos anciennes provinces. Il cumulait aussi des revenus considérables, qu'il a bien gérés et qui lui permirent d'exercer un mécénat entre autres en faveur des membres de sa famille et de l'Université de Louvain. Après le décès de sa femme Jacqueline Damant (en 1552), il s'orienta également vers les dignités écclésiastiques. En 1562, il fut ordonné prêtre et il obtint de nombreuses fonctions écclésiastiques, notamment celle de prévôt du chapitre de Saint Bavon à Gand, dignité à laquelle étaient attachés également de très gros revenus.[5)]

Il n'entre nullement dans nos intentions d'étudier en détail la carrière de Viglius. Tout au plus voulons nous appeler l'attention sur certaines fonctions qu'il a été amené à exercer, notamment celle de bibliothécaire et qui nous montrent qu'il avait une érudition remarquable pour son époque. Ensuite nous insisterons sur son intérêt pour les cartes.

Viglius bibliothécaire

Par une ordonnance datée de Bruxelles, le 12 avril 1559, Philippe II procéda à la création à Bruxelles d'"une belle librairie ou bibliotecque". Il chargea Viglius de Zuichem "de rassembler tous les livres estans en nos pays de pardeça à nous appertenans". Il s'agissait de récupérer et de regrouper en premier lieu, tous les livres ayant appartunu à Marie de Hongrie et de les réunir aux collections du roi qui se trouvaient éparpillées dans les différentes résidences du prince dans les Pays-Bas, que ce soit à Lille, à La Haye, à Bruxelles ou ailleurs.

Peu de temps après, Corneille Breydel, secrétaire du prévôt de Saint Bavon à Gand et un de ses exécuteurs testamentaires, désigna Viglius sous le titre de préfet de la bibliothèque royale in "aula Bruxellensi".[6)]

Le président du Conseil Privé n'était pas seulement chargé de regrouper, de garder et de conserver la bibliothèque du roi; il devait également s'occuper d'agrandir les collections "l'augmenter de tels livres qu'il verra convenir".[7]

Philippe II appréciait la grande érudition, la formation scientifique approfondie du nouveau bibliothécaire, ses relations avec un nombre imposant de savants et d'humanistes de son temps, comme on peut encore s'en rendre compte, entre autres par sa correspondence fort étendue.[8]

Le premier bibliothécaire avait toutes les qualités pour regrouper, gérer et enrichir l'importante collection de livres manuscrits et imprimés créeé jadis par les ducs de Bourgogne. Viglius commença sa mission en récupérant assez rapidement la bibliothèque que la gouvernante Marie de Hongrie avait installée dans son château de Turnhout en 1555-1556.[9]
Sa tâche fut facilitée par une stipulation de l'acte du 12 avril 1559, où Philippe II ordonna à ses conseillers et gardes des chartes et joyaux de faire parvenir à son bibliothècaire les inventaires des livres qu'ils détenaient et ensuite les ouvrages que celui-ci leur réclamerait. En contre-partie, Viglius était obligé de leur délivrer un reçu de tous les ouvrages envoyés.[10] Déjà le 22 mai 1559, le nouveau bibliothécaire avait récupéré les 333 livres de la bibliothèque de Marie de Hongrie, mentionnés dans un inventaire dressé antérieurement. Sur ce même document, il notifia la décharge officielle.[11] Cette tradition de dénombrer et de vérifier les fonds de livres et autres documents au moyen de répertoires bibliographiques ou inventaires, Viglius la poursuivit jusqu'à la fin de ses jours. Plusieurs de ces inventaires ayant été conservés en original ou en copie, ils rendent encore actuellement beaucoup de service aux historiens des collections de livres de nos anciens Pays-Bas, parmi lesquelles la Bibliothèque de Bourgogne.

En 1572, Viglius fit rédiger un catalogue de la bibliothèque de la cathédrale de Saint-Bavon à Gand. Ce catalogue qui a été conservé, a permis à notre collègue Prof.Dr. K.G. Van Acker de Gand de trouver de précieux renseignements sur la belle et importante collection de manuscrits que l'abbé de Saint-Bavon Raphaël de Mercatel († Bruges, 3 août 1508) s'était constituée à la fin du XVe et au début du XVIe siècle.[13)]

En 1569, le mécène frison fonda le Collège Viglius à Louvain. 12 bourses y étaient créées et réservées à des jeunes gens pauvres qui voulaient devenir prêtres. 6 étaient destinées à des Frisons, les 6 autres à des candidats choisis parmi les enfants de choeur de la cathédrale de Saint-Bavon à Gand. Viglius légua tous les livres imprimés de sa bibliothèque à ce collège qui possédait une bibliothèque dès ses députs. Les statuts (art. 17 et 39) étaient rigoureux quant à la consultation des livres et à leur restitution rapide.[14)]

Viglius et la cartographie

Nous n'approfondirons pas le rôle de Viglius comme bibliothécaire pour traiter de son importance pour les cartes et leur histoire.

Comme il avait peu l'occasion de se rendre en Frise où il avait des biens et des revenus considérables, il a fait dresser des cartes et des plans de ses terres et des bâtiments qui s'y trouvaient. Ainsi, il avait ses propriétés devant les yeux. Il fit la même chose pour les domaines qui lui procuraient ses revenus comme prévôt du chapitre de Saint-Bavon à Gand.[15)]

Au Grand Conseil de Malines existait déjà vers le milieu du XVe siècle une tradition de faire dresser des cartes et des plans comme pièces justificatives pour certains procès.[16)] Ainsi Viglius pouvait y développer son intérêt, son goût pour les cartes. En réalité le juriste frison s'intéressait à ces documents au moins depuis 1534, ainsi qu'il en a fourni lui-même le témoignage.

Le 18 septembre de cette année [17)] il écrivait à son ami Adrianus Marius (Malines - Bruxelles, 1568), le frère de Janus Secundus pour le remercier de lui avoir envoyé une carte. Toutefois celle-ci ne lui donnait pas satisfaction malgré toute la peine donnée. Il avait, en effet, une carte, probablement dressée par un marin, qui n'était pas assez correcte pour qu'il ose la présenter à Cranevelt (Nijmegen,

1485-Malines 1564). Il fait allusion aussi à une carte du diocèse d'Utrecht où étaient représentés tous les lieux ainsi que les distances, les espaces, les sources des fleuves, leurs cours et leurs embouchures. Il s'est souvenu alors qu'il avait vu jadis dans la maison de Wilhelmus Goudanus à La Haye une très grande carte, qui, sauf erreur, était imprimée. Comme à ce moment, il ne s'intéressait pas spécialement à ce document, il ne se souvenait plus comment était cette carte ni où elle avait été publiée. Il insiste auprès de son ami pour qu'il l'informe au sujet de toutes les cartes de ce genre, de leur prix et comment il pourrait les acquérir. Il exprime l'espoir que cette partie des Pays-Bas insuffisamment appréciée et connue (ce qui n'est pas les cas de la vôtre qui est tournée du côté de la France) sera bientôt représentée de la même manière.

Cette lettre au poète Adrianus Marius nous fournit un témoignage formel, quoi qu'un peu énigmatique, sur des cartes des Pays-Bas qui existaient déjà en septembre 1534 et même avant cette date, ainsi que sur d'autres qui étaient en cours d'exécution. Faut-il y voir une allusion aux travaux préparatoires de Jacques de Deventer qui à ce moment s'occupait probablement déjà des levers pour ses cartes des provinces des Pays-Bas et notamment pour celle du duché de Brabant. N'est-ce pas à cette carte dont on sait qu'elle fut présentée antérieurement au 20 décembre 1536 au Conseil du Brabant à Bruxelles, que Viglius fait allusion?[18)]

Nous pouvons ainsi admettre comme date probable des débuts de la collection de cartes de Viglius l'année 1534. Il serait sans doute possible en étudiant attentivement sa correspondance de connaître l'un ou l'autre détail sur les acquisitions faites par le savant collectionneur au cours des années, mais nous ne sommes pas assez familiarisé avec la documentation relative au président du Conseil Privé et du Conseil d'Etat pour nous y risquer. D'ailleurs, cela nous conduirait à des développements dépassant le but proposé dans ces pages. Nous nous contenterons de traiter du catalogue de ses cartes dressé en août 1575. Mais avant cela, nous rappelons que Viglius recourait aussi aux services du géomètre-cartographe gantois François Horenbaut. Il a payé les études universitaires à Louvain de Joannes Horunbaut, fils de celui-ci, en l'admettant comme "alumnus et Flandria" au Collège Viglius jusqu'au 25 février 1576 quand il renonça à la carrière ecclésiastique.[19] Il eut à son service Virgilius Gheys qui a dressé des cartes pour lui.

L'histoire de la cartographie et la connaissance de celle-ci doit aussi au président du Conseil Privé et prévôt du chapitre de Saint-Bavon à Gand, le sauvetage, la conservation et l'identification de l'unique collection de plans de villes des Pays-Bas dressée par Jacques de Deventer (c. 1505-Cologne 1575) entre c. 1540 et 1575. Virgilius Gheys intervint dans cette affaire.

Dans le compte des dépenses de Viglius pour l'année 1575, il est marqué à la date du 11 mai, une dépense de "12 Philips daelers" pour le serviteur du châtelain de Gouda qui apporta la nouvelle de la mort de Jacques de Deventer. Le 3 juin, un payement est fait à Virgilius Gheys, à propos des recueils de plans que J. de Deventer avait dressés pour le roi. Le 15 octobre, 14 Philips daelers, sont payés au messager de Cologne qui avait apporté les "charte boecken" de J. de Deventer.[20]

Ce document comptable nous apprend que Jacques de Deventer est mort avant le 11 mai 1575 puisque Viglius en eut connaissance à Bruxelles à cette date. Nous ignorons quelles furent les prestations de Virgilius Gheys quant au recueil de plans de Deventer (au début de juin). S'est-il rendu à Cologne sur ordre de Viglius? Finalement, nous apprenons ainsi que le recueil de plans que Deventer avait avec lui à Cologne au moment de sa mort, était déjà restiqué à l'administration des Pays-Bas à Bruxelles le 15 octobre 1575. Ces plans, fruit de longs et coûteux travaux sur le terrain, étaient des documents secrets d'une grande importance militaire et politique. Comme haut dignitaire et serviteur consciencieux de son roi, malgré son état de santé défaillant, Viglius a tout mis en oeuvre pour les récupérer et les faire envoyer en Espagne. C'est là qu'on a retrouvé au XIXe siècle deux des trois volumes des plans de villes, mis au net, de Jacques de Deventer.

Viglius avait fait confisquer ces plans et les mettre en lieu sûr par le magistrat de Cologne, tout de suite après la mort de Deventer (vers le 11 mai 1575). Il resta en rapport étroit avec les édiles de Cologne jusqu'à la restitution des trois volumes qui furent envoyés à Bruxelles chez Viglius et arrivèrent en Espagne après sa mort.[21)]

Entretemps, au mois de novembre 1575, Viglius avait eu connaissance de la présence à Malines de Barbara Smets, la compagne de Deventer, et des minutes des plans de ville du grand topographe qu'elle détenait. Il en fit part à son ami frison Joachim Hopperus, qui s'occupait à ce moment à Madrid des affaires des Pays-Bas, le 23 novembre 1575, et lui écrivit qu'il mettait tout en oeuvre pour récupérer des documents cartographiques en promettant une sérieuse récompense à Barbara Smets.

D'autres précisions sur le sort de ces minutes nous manquent jusqu'au moment où une partie en réapparaît en 1859, lorsqu'elles furent présentées lors d'une vente publique chez Martinus Nijhoff. Ces documents anonymes furent finalement identifiés et prirent leur place dans l'histoire de la cartographie grâce à la correspondence échangée entre Viglius et Hopperus. Celle-ci avait été publiée par C.P. Hoyinck van Papendrecht à La Haye en 1743, dans ses *Analecta Belgica* et à Utrecht en 1802. P.C. Molhuysen, archiviste et bibliothécaire à Deventer, appela dès 1839 l'attention sur cette correspondance. Ce ne fut toutefois qu'on 1866 que Frederik Muller et Eekhof parvinrent à identifier définitivement ces minutes des plans de ville comme l'oeuvre de Jacques de Deventer.[22)]

Il nous reste à traiter du catalogue des cartes du savant mécène frison.

La collection de cartes de Viglius

Depuis que de 18 septembre 1534 il avait témoigné son intérêt pour de bonnes cartes des Pays-Bas jusqu'en août 1575, Viglius s'était constitué une riche collection de cartes. A ce moment, il en fit dresser le catalogue ou plutôt une liste sommaire. Celle-ci a été conservée et publiée en 1863 par A. Pinchart. Il légua cette importante série de documents géographiques au Collège Viglius à Louvain. Voici le titre: *Regionum, locorumque descriptiones, seu, ut vulgo vocant, chartarum catalogus, secundum ipsorum situationem, conscriptus mense augusto 1775.*[23] Cet ensemble de documents cartographiques, plus de 220 au total, comprend 5 cartes du monde dont une publiée à Zurich en 1534 et celle de Gaspard Vopel, de grand format, imprimée en 1549. La carte du monde de Gérard Mercator publiée en 1569 y figure ainsi que celle d'Oronce Fine publiée à Paris en 1536; une autre, dont l'auteur n'est pas cité, fut publiée à Paris en 1547. Il y a deux "chartes de la grande mer".

L'*Europae descriptio* de Gérard Mercator (il n'est pas indiqué si c'est l'édition de 1554 ou de 1572) était pendue près de la cheminée de la bibliothèque.

Une carte de l'Asie de 1567, une du Canada (Nova Francia) publiée à Venise en 1566 et à la fin de l'inventaire, une "urbs Africa bis", trois parties de l'Asie, gravées, ainsi qu'une carte de l'Egypte éditée par Ortelius, voilà toutes les cartes des continents non européens, auxquelles il faut ajouter 4 cartes de la Terre Sainte.

Nous trouvons surtout des cartes des pays de l'Europe et des cartes partielles de ces pays: 2 cartes de l'Espagne, 1548 et 1571; 3 de la Bretagne, 1 d'Irlande, 6 de la France, dont une publiée par A. Nicolai à Anvers et une par Gérard de Jode en 1568. 6 cartes de parties de la France, non spécifiées, suivies de 18 cartes des provinces, des régions et villes de ce pays dont le nom est cité. Après une Gallia Belgica publiée à Anvers par de Jode en 1566, suivent des cartes et des plans de l'Artois, du siège de Tournai, du Hainaut, du village de Faulx.

Après la carte de Flandre de Mercator, on trouve un grand plan de Gand de 1540, un petit plan de Bruges de 1563. Il y a deux cartes de biens de Saint-Bavon: les "meerschen van Wolfsdonck" et le "Lereghem-Velt".

La carte du Brabant "Brabantiae descriptio" de J. de Deventer de 1536 est suivie des plans d'Anvers, de Bruxelles, de la carte de la Zélande de J. de Deventer (sans indication de date) et d'une édition plus récente, plusieurs cartes partielles de la Zélande.

Een groote chaerte van Hollandt, imprimée en 1542 (est-ce celle de Deventer?) précède plusieurs cartes partielles de la Hollande et d'autres parties des Pays-Bas septentrionaux dont une du siège de Leyde de 1574. Une carte de la Merwede, des plans d'Amsterdam, de Texel et de plusieurs territoires voisins. La collection est riche en cartes de Frise: celle par Deventer de 1545, une autre manuscrite, une partie de la Frise, des cartes de villes et villages de la Frise, dont au moins 3 de Zwichem. Citons encore la carte du duché de Gueldre de J. de Deventer, 1545, une autre par Chr. Sgrooten, éditée par H.Cock et une carte manuscrite par le même Sgrooten.

Nous ne pouvons que mentionner les cartes du diocèse de Liège, de la Moselle, du Luxembourg, de Jülich, Berg, Clève avec la province de Cologne par Sgrooten, St. Hubert, Metz, le comté de Namur, les duchés de Luxembourg et de Lorraine. Une carte de Westphalie copiée de Sgrooten par Virgilius Gheys; une autre est de Sgrooten. Suivent des cartes et plans de parties de la Westphalie, des Pays-Bas de l'autre côté du Rhin, de Paderborn, Linghen, de Münster et Overijssel ainsi que d'autres parties de l'Allemagne. Les cartes de ce pays sont très nombreuses.

Du duché d'Autriche, il y a 4 cartes; de la Hongrie 3 dont 1 de Gastaldi publiée en 1546.

Outre la Dalmatie, il y a des cartes de l'Italie et quelques unes de parties de ce pays.

L'inventaire, tel qu'il est publié par Pinchart, se termine par 6 cartes, dont celle des Pays-Bas, destinées au duc de Savoye comme cartons pour des tapisseries.

Ce répertoire des cartes de Viglius mériterait une étude approfondie comme ce fut le cas pour le *Catalogus auctorum* des premières éditions du *Theatrum Orbis Terrarum* d'Ortelius[24]. Toutefois, il est plus sommaire et moins précis.

En traitant de Viglius, nous avons appelé l'attention sur ses études, son érudition, ses relations, ses hautes dignités, parmi lesquelles celle de bibliothécaire de Philippe II.

Son rôle comme amateur et collectionneur de cartes, ainsi que dans la restitution et le sauvetage des plans de villes des anciens Pays-Bas par J. de Deventer, lui valent une place dans l'histoire de la cartographie. Viglius mérite d'être mieux étudié.[25]

NOTES

1) S.P. HAAK: *Aytta (Wigle) van Zwichem. In: Nieuw Nederlandsch Biografisch Woordenboek, 2, Leiden, 1912, col. 46-52.* H. DE VOCHT: *Literae virorum eruditorum ad Franciscum Craneveldium 1522-1528. Louvain, 1928 (Humanistica Lovaniensia, 1), p.690-692.* H. DE VOCHT: *History of the Collegium Trilingue Lovaniense, 1517-1550, Louvain, 1951-1955, 4 vol. (Humanistica Lovaniensia, 10-13.* L. ROBYNS et J. MERTENS: *Viglius Aytta van Zwichem. In: 550 jaar Universiteit Leuven, Catalogus-tentoonstelling, 2, Leuven, 1976, p.295-296.*

2) M. BAELDE: *De collaterale raden onder Karel V en Filips II (1531-1578), Brussel, 1965, (Verhandelingen van de Koninklijke Academie van België, Klasse der Letteren, 60), spécialement p.324-325.* R. DESMED: *Une version inédite de la biographie latine de Viglius D'Aytta. In: Archives, Bibliothèques et Musées de Belgique, 30, (1959), p.266-274.* J. LAMBERT: *Viglius van Aytta et Erasme. In: Miscellanea historica Alberti De Meyer, Louvain, 1946, p.812-824.* E.H. WATERBOLK: *Viglius van Aytta in gesprek met Hieronymus Frobenius. In: Flecht op 'e koai, Groningen, Wolters-Noordhoff, p.133-141.*

3) L. GALESLOOT: *Everardi (Nicolas). In: Biographie Nationale, 6, Bruxelles, 1878, col. 751-754.* J. VAN KUYK: *Everardi (Nicolaus) de Middelburgo. In: Nieuw Nederl. Biogr. Wdb., 3, Leiden, 1914, col. 358-360.* H. DE VOCHT: *Literae ... ad Craneveldium, p.336-338. In: 550 jaar Universiteit Leuven, Catalogus, 2, p.294-295.*

4) M. BAELDE: *Op, cit., p.324-325 et passim.*

6) LAMBERT: *La Bibliothèque royale*, p.237.

7) CL. LEMAIRE *et* M. DEBAE: *Esquisse historique 1559-1837. In: Bibliothèque Royale, Mémorial, Bruxelles, 1969, p.3 ss.* LAMBERT: *La Bibliothèque royale.*

8) J. VAN DEN GHEYN: *Catalogue des manuscrits de la Bibliothèque royale de Belgique, vol. 7, p.463-468.* H. DE VOCHT: *Literae ... ad Fr. Craneveldium, p.690-691.* ALLEN: *Opus epistolarum D. Erasmi Roterodami, t. 8, Oxford, 1934, p.56-57.* LAMBERT: *La fondation du Collège de Viglius, p.711.*

9) LAMBERT: *La Bibliothèque royale, p.239.*

10) LAMBERT: *Op. cit., p.238 et note 2.*

11) LAMBERT: *Op. cit., p.239.*

12) LEMAIRE *et* DEBAE: *Mémorial, p.3-4. Cfr. notre note 23.*

13) *Viglius en de Librije van Sint-Baafs te Gent. In: De Gulden Passer, 39, (1961), p.339-357; paru également in: Handschrift en Druk, Lezingen gehouden in ... het 24ste Filologencongres te Leuven (6-8 april 1961), p. 85-103.* VAN ACKER: *De Librije van Raphael de Mercatellis, abt van Sint-Baafs en bisschop van Rhosus. In: Archives et Bibliothèques de Belgique, 48, (1977), p.143-198. Cfr. Marcatellis (Raphaël de), notice par J. Frederichs. In: Biographie Nationale, 13, (Bruxelles 1894-1895), col. 420-423.*

14) J. LAMBERT: *La fondation du collège de Viglius. Vigliana, Groningen 1975, p.99-123.*

15) *Viglians, Groningen, 1975, p.1-3, 17 et passim.*

16) A.H. HUUSSEN jr.: *Jurisprudentie en kartografie in de XVe en XVIe eeuw, Brussel, Algemeen Rijksarchief, 1974 (Miscellanea Archivistica, V).*

17) *Lettre à Hadrianus Marius, du 18 septembre 1534, à la Bibliothèque Royale à Bruxelles, Ms. II 1040, Viglii Epistolae, t. I, f° 92-94: Memini et olim in aedibus Wilhelmi Goudani in Haga Comitis quandam me vidisse amplissimam mappam, et ni fallor, impressam. Sed tunc eius studio non tenebar, ut qualisnam illa fuerit, aut ubi impressa non recorder. Ce document est signalé in Vigliana, Groningen, 1975, p.13 et note 35. Sur W. Goudanus (c. 1492-6 décembre 1555), voir* ALLEN, *Opus epistolarum D. Erasmi, V, n° 1320, p.138 et passim.*

18) B. VAN'T HOFF: *De kaarten van de Nederlandsche provincien in de zestiende eeuw door J. van Deventer. S'-Gravenhage, 1941, inleiding, p.5, 11.* A. DE SMET: *Mechelse geleerden op het gebied van wiskunde en kartografie. In: Album Antoine De Smet. Édité par le Centre national d'histoire des sciences sous la direction de Lisette Danckaert, Bruxelles, 1974, p.167-169.* A. DE SMET: *La cartographie hollandaise, Bruxelles, 1971, p.9-16. Nous y avons insisté sur une priorité de près de six ans de Deventer par rapport à Gemma Frisius en ce qui concerne leurs études à Louvain (Deventer, immatriculé le 24 avril 1520, Gemma, le 26 février 1526. Ce détail était inconnu de Van't Hoff.)*

19) *Sextus (alumnus ex Flandria) Joannes Horenbaut Gandensis promotus 1576 qui eodem anno promotionis 25 februarii discessit quia sacris initiari cum tempore noluit. Vigliana, p.117.*

20) *Vigliana, p.22, Meye: Opten XI^en dach gegeven Willem dienaere vanden Castellein vande Goude brynghende de tydinghe van toverlyden van m^r Jacob van Deventer XII philips daelers maekende ...*
Junius: Opten III^en dach gegeven Virgilius Gheys ter causen vande boecken die m^r Jacob van Deventer gemaeckt hadde voerden Coninck ...
p.24, October: Opten XV^en dach heeft den bode van Coelen die charte boecken van m^r Jacob van Deventer ghebrocht ende heb hem gegeven XIIII Philips daelers ...

21) B. VAN'T HOFF: *De Nederlandsche gewestelijke kaarten en stedeplattegronden vervaardigd door Jacobus van Deventer. In: Nederlandsch Archievenblad, 47, (1939-1940), p.19-32.* B. VAN'T HOFF: *Jacob van Deventer keizerlijkkoninklijk geograaf. 'S-Gravenhage, 1953, p.20-23 et spécialement les documents n°s 20 à 39, p.41-50. Cfr. notre note 20.*

22) VAN'T HOFF: *De Nederlandsche gewestelijke kaarten en stedeplattegronden, p.26-29.* VAN'T HOFF: *J. v. Deventer, 1953, p.26-28 et doc. 34, p.47-48.*

23) A. PINCHART: *Collection de cartes de Viglius. In: Archives des arts, sciences et lettres, 1ère série, t. 2, Gand, 1863, p.311-315. Réimprimé in: Imago Mundi, V, 1948, p.18-20. L'original a été retrouvé par* E.H. WATERBOLK, *Viglius of Aytta, Sixteenth Century Map Collector, in: Imago Mundi, 29, 1977, p.47 et note 10 (Archives Générales du Royaume, Bruxelles, Inventaires de la 2e série, nº 419, fºs 140-149.)*

24) L. BAGROW: *A. Ortelii catalogus cartographorum, 2 vol., Gotha, 1928-1930.*

25) *Nous tenons à remercier tous ceux qui nous ont aidé à la Bibliothèque royale à Bruxelles et au Centre national d'histoire des sciences, spécialement Mme Andrée Menu - Van Loo, secrétaire de ce Centre.*

THE DUTCH UNION MAP CATALOGUE
FROM RESEARCH PROJECT TO OPERATIONAL PROJECT

By E. Hans van de Waal

In his Ph.D. thesis, called "MARC-Maps, the history of its development and a current assessment", [1] Hugo Stibbe particulary studies the various bibliographical aspects connected with the application of a MARC (Machine Readable Catalogue) system to the description of maps. The MARC system has been developed at the Library of Congress in Washington, D.C. In this context Stibbe also deals with the CCK (Centrale Catalogus Kartografie), in English the DUMC (Dutch Union Map Catalogue). In this article I will explain the function of the DUMC within the Dutch information system. The research phase of the project DUMC will come to an end in 1978. After this the emphasis will be on its functioning as an operational union catalogue for cartographic documents. From the beginning of 1976, when Stibbe finished his research, some new developments occurred in the DUMC project, of which I will give a synopsis.

The Initial Stage

The establishment of the DUMC had its origins in a series of coincidental occurences at the end of the 1960s. There was a growing belief that it would be impossible to cope with the increasing supply and demand in the information systems, unless close co-operation developed between all Dutch libraries and map collections, and automation was applied. This resulted in the foundation of the Netherlands Organization for Information Policy (NOBIN). For its budget this institution came under the Ministry of Education and Sciences and the Ministry of Economic Affairs. Its management was in the hands of persons from institutions directly concerned with information policy, both actively and passively. NOBIN was granted the disposal of a budget for the stimulation of developmental and research projects.

In 1968 a working group DUMC had been formed, with Professor Dr. Ir. C. Koeman as its Chairman. The working group's first initiative was to investigate the size and state of the Dutch map collections. The enquiry aimed to collect data that would be of interest for the realization of a DUMC. It asked questions about the number of maps, atlases and aerial photographs held, and also the kind of maps, catalogues and staffing, and it invited ideas and opinions about what a DUMC should be. Compared to the book collections, the map collections appeared to be a deplorable condition in terms of arrangement, conservation and cataloguing. There were various reasons for this. To maps in libraries we could apply the Dutch saying, "unknown, unloved". Through lacking distinctive bibliographical data such as title and author, maps evidently have failed to command the interest of librarians, some of whom also are put off by the mathematical and technical considerations involved in the study and handling of maps.

Most map collections were found in libraries and archives, where they were being treated as marginal material. Those directly responsible for the map collections were trained as librarians and knew little about maps themselves. The enquiry's results, however, indicated a growing interest in maps on the part of the users and of the controlling institution, and showed that most map collections supported the idea of a DUMC.

In 1970 the DUMC working group gained official status and became the Nederlandse Stichting Informatie- en Documentatiecentrum voor de Kartografie, NSIDK (Dutch Foundation Information and Documentation Centre for Cartography). All large map collections and some of the institutions concerned are represented. On the general board sit representatives from the archives, the libraries and the map-producers, and from institutions such as the land registry and the Royal Academy for Science.

The foundation's main activity is to administer the DUMC Bureau, entrusted with the development of a union catalogue for maps. The Ministry of Education and Sciences granted DUMC a subsidy from which the staff's salary could be paid. From 1971 to 1975 the grant was paid through NOBIN, which also supervised the management of the project. Since 1976 the subsidy goes from the Ministry directly to the DUMC, but NOBIN still has an advisory task, towards the Ministry as well as towards DUMC.

The project became a research project of the cartographic department of the Geographical Institute of the State University in Utrecht. The University contributes a part of the work-time of the director, who is employed at the cartographic department, and provides rooms and computer facilities. The very close co-operation with ACCU (Academic Computer Centre Utrecht), which developed the computer programmes, proved to be essential to the realization of the project.

Co-operation between the DUMC working group and the Geography and Map Division of the Library of Congress was facilitated through the personal association of Professor Koeman of the Geographical Institute and Dr. W.W. Ristow, Chief of the Geography and Map Division, L.C. In the later months of 1969 we learnt that a MARC Data Preparation Manual was being prepared. We were able to obtain a Xerox copy of this Manual, which was very useful to us when we were formulating the project's objective and procedure, with a view to applying for a subsidy from NOBIN. DUMC chose to work as much as possible along the lines of MARC-Maps as developed at the Library of Congress. It was concerned only to make those alterations necessary in the Dutch context to describe cartographic documents in such a way as to provide access to elements such as region, subject, scale, author and co-ordinates. It seemed undesirable to develop a new system of our own, because of the lack of necessary expertise, the very limited means and the advantages of international co-operation which we aimed for from the outset.

In view of the present-day development of international co-operation in the field of bibliography, it is necessary to stress the importance of keeping all parties informed of decisions and plans before a publication, which is to serve as a manual for co-operation, reaches its final or nearly final stage. This could be arranged by provisional publications, concerned with announcing aims, procedures and contents. This would give other institutions the opportunity to take measures in good time. Criticism and suggestions may be called forth, but we remain free to take them into account, to say yes or no. With complicated co-operational groups it is impossible to reply on personal relations between the members. The play of coincidence and the personal character of the participants would then assume too great a role in the developments. Inside information from the Geography and Map Division of the Library of Congress and from the MARC Office in Washington saved us wasting time on problems that were already being solved there. It would be useful if this kind of inside information became more generally available by means of provisional publications.

Objectives

DUMC's aim and procedure have been summarized in 1972 in an evaluation report (in stencil), from which I quote:

1) The DUMC project's objective:

   To make Dutch maps, map series, atlases, maps in books and periodicals accessible via. a data bank, consisting of codes descriptive elements. With the aid of this data bank it will be possible:

   - to print catalogue cards and complete catalogues on various subjects and regions for the various map collections in Holland,
   - to make a selection of descriptive elements relevant to maps and to sort by combinations thereof, including region, scale, title, date, coordinates etc.,
   - to exchange, by means of magnetic tapes, bibliographic data with foreign "DUMC"-projects that are also based on the MARC system. The United States already have a MARC system specially adapted to maps. In Canada such a system is being worked out. It may be expected that countries developing a national MARC will also include documentation of maps in their system.

2) DUMC's raison d'être and benefits:

   A central effort is needed because:

   - knowledge of maps and of map documentation is

scarce,

- automated procedure, because of the computerized handling, calls for an instrumentation that is only workable in a central approach,
- most individual collections do not have the means to catalogue their maps properly and they cannot cope with the growing demand for maps as sources of information.

The central approach has several advantages:

- DUMC created a procedure for abridged descriptions, in which only the deviations from maps already catalogued need to be stated. This prevents duplication of work, as many maps occur in more than one collection and many have only minor differences one from another. In this way libraries and archives can take part in DUMC quite simply and at an early stage. The maps can be entered under an individual number, and the controlling institution thereafter can have at its disposal output adapted to its own wishes.
- Because of international contacts, the various national union catalogues can avoid duplication of work.
- The map collections associated with DUMC at home and abroad are being opened up and becoming more easily accessible. Also the maps in books and periodicals, now almost unattainable, will be accessible, once their bibliographical records are entered into the DUMC data base.

Procedure

The pilot project DUMC lasts from 1969 to 1979. In 1973 the project's first phase came to an end. During this phase the system had been developed and some five thousand maps had been catalogued. First of all the MARC Format for Maps and the "Dutch Directives for the Bibliographic Description of Maps" had to be adapted to one another and to the special characteristics of maps. The ACCU developed programmes to put the descriptions on magnetic tapes and also programmes for the additional mutation and correction procedures. From the already created data base, catalogue cards and catalogues were printed. It was essential to provide the facilities to sort by scale and co-ordinates, by the dating of the map's contents and by other elements. It is not necessary to explain why the limitless number of subjects to be mapped made it undesirable to draw up a special system for the subjects shown on maps. We have chosen the UDC for the regional as well as for the systematic classification. The regional classification can be extented by means of keywords, to be chosen by the local cataloguers themselves. After experience had been gained during cataloguing, we drafted user-lists on subject and region, to make it unnecessary constantly to turn to the UDC handbooks. A great deal of independence is left to the local map collections in their actual cataloguing procedures. Matters such as the choice of main entry and the print-out of catalogues or catalogue cards can be decided by the local collections on their own, and they may follow a practice which deviates from that of the central system if financial constraints permit this. Apart from the DUMC classifications, individual classifications can be adopted. At the end of this article, examples are given of the present-day bibliographic techniques and print-outs on main entry (region), subject, title and author.

In the period from 1974 to 1979 DUMC has the task of cataloguing as many maps as possible from the collection of modern and early maps belonging to the Geographical Institute in Utrecht. Thus a data bank will be established, to which other collections can turn when cataloguing their own maps. This is made possible by taking automatically from other maps the descriptive elements with identical texts, the so-called Mother-Daughter procedure. For duplicates we only need to give the call number in the local map collection and a reference to a previously made description; after this the computer duplicates the whole description and is able to deliver the desired print-out. In this period we shall also make "authority lists" for names of authors, series and titles, as these can cause much confusion for map cataloguers.

While establishing DUMC, we took it for granted that the local map collections would make their own descriptions. The DUMC Bureau would take care of the assimilation of the data into the central data base. In this data base all the information from the various collections has been assembled. In effect, each associated collection has its own data base, which is accessible to all.

Several misunderstandings have arisen over this method of taking over descriptive elements by using a simple reference. Though the procedure is used extensively with maps of the same series, it has nothing to do with the ISBD's multilevel descriptions. In principle, the Mother-Daughter procedure can be used in any example where maps have some descriptive elements in common. For identical maps in different collections, for instance, we often add a call number only. This method is especially useful for maps, both because maps so often run to many editions and because they often belong to series or sets.

New Developments

When the project started, we decided that the computer-programmes to be developed should be as simple as possible, partly because of the limited financial resources. In 1973, when we learned that the Academisch Computer Centrum te Utrecht (ACCU) was going to change from an IBM X8 computer to a Cyber 72, we chose to give the data base a MARC structure, one which would permit us to generate in a simple way full-grown MARC tapes with a leader and a record directory. This development became possible because the ACCU started to provide computer services for similar projects, to the benefit of DUMC. Moreover, the size of the data base gradually made a more sophisticated computer approach necessary. It had already become clear that DUMC was able with difficulty to use elements of MARC Map tapes from the Library of Congress. To be able to produce MARC tapes oneself was advantageous too, in the context of shared cataloguing and international co-operation. The establishment of ISBD(CM) and of INTERMARC and UNIMARC (which still have to be worked out for maps), gives reason for optimism.

Another important step forward, one which extends the system, is the automation of authority lists for UDC-region (and keyword), UDC-subject, name and series title. The use of authorities, by which certain elements in a standardized form are taken into a map description, is not generally accepted in Holland. In most cases author and title are rendered as they occur in the bibliographic entity, a moderate use being made of 'see' and cross-references. As authors of cartographic documents and titles of map series generally are not mentioned clearly on the documents, it is necessary to standardize. Otherwise, a confused situation with too many added entries and references would arise. The system used for the creation of authority files resembles that of the Canadian MARC, as described by H. Stibbe in his thesis. It would be advantageous if UNIMARC also were to use authority lists. They could then use the system developed in Canada, where language problems are avoided by coding the authorities. Another attractive aspect of the Canadian MARC is the possibility of using a code which makes it possible to receive the text in the desired original language from the authority files, by means of machine interface between authority file and bibliographic file.

From the beginning of the pilot project we had to deal with photomaps and this made us wonder if it would not be possible to catalogue all topographic documents on the same system and to bring them together into the same data base. This appeared to be possible, if we applied to cartographic documents a definition identical to the one that the ISBD(CM) uses for "cartographic material": "... all materials representing, in whole or in part, the earth or any celestical body at any scale." This means that separate maps, map series and maps in periodicals, as well as aerial photographs, bird's eye views and topographic prints can be collected in the same data base with the same rules. It is possible to move even a step further and to develop a means to catalogue potentialities for producing maps. At the moment the UNIMARC committee of the IFLA Working Group ISBD(CM) is working on this. They are considering cataloguing printing plates, aerial photograph negatives, the procedures for making computer maps, and data bases of remote sensing techniques such as Multi-spectral Scanning, and they will do this by treating these information suppliers as documents. Notably with these last two techniques, it is more important to know what possibilities in various locations are available for producing cartographic documents by automation than to trace the documents themselves. These, moreover, would be much more difficult to handle because of their great variety and the fact that they are made one by one.

Subsidiary Effects

The DUMC Bureau has contributed indirectly to the enhanced status of the map as a document. Because of this modern approach the project gave maps a place of their own, next to books and non-books, both of which are going through the same developments in documentation. The ISBD(CM) has played an important and positive part in all this. The NSIDK brings together various institutions concerned with maps. Now that land registries and university libraries realize that they share common problems in the field of cartobibliography, a sense of solidarity is growing, uniting the interests of all who are concerned with maps. Because of this solidarity it was possible to start a training course for map librarians. The NSIDK brought about the establishment of a course for map librarians during the specialization year at a Higher Training College for archivists and librarians. Those already working in a map collection are offered a course of thirty hours on the use and application for cartographic documents at the cartographic department of the Geographical Institute (State University Utrecht); at the end of the course the NSIDK awards a testimonial. The training course is very important to DUMC, because a central catalogue based on locally made descriptions cannot function unless local map librarians have a sound knowledge of map cataloguing.

The relations within the board of the NSIDK and between NSIDK and the Dutch Society for Cartography, of which among others Dutch map producers are members, made it possible for the Koninklijke Bibliotheek (Royal Library) to take advantage of the system of "Vrijwillig Depot van Nederlandse Publicaties" (Volontary Deposit of Dutch Publications). For this reason a map cataloguer of the Royal Library has been appointed to the DUMC Bureau, and makes descriptions of all newly issued maps. From these descriptions we are able to publish with the help of the computer a bibliography of maps published in Holland. The first number of this work, entitled *Bibliografie van in Nederland verschenen kaarten*, comprising the output for the year 1975, was published in September 1977. The bibliography for 1976 will be published in the spring of 1978. We intend to publish each spring the bibliography of the previous year.

The DUMC bureau also helped to found in 1975 the working group on Map Curatorship within the Dutch Society for Cartography. This working group is still too young to have acquired a definite character, but a great number of interested members attended the meetings about such subjects as the marginal information on maps and map conservation.

The Future beyond 1978

Before I go into the DUMC developments after 1978, when the emphasis will shift from research to functioning as an operational project, I will give some background information on information policy in Holland. Public institutions in the Netherlands are remarkable for their decentralized structure. Holland originated from a municipal-civic culture which can be felt even today. Municipalities in the Dutch governmental system have a great deal of autonomy when compared to those in other countries. The field of librarianship also reflects this decentralization and autonomy. For instance, Holland does not have a real national library, though the Royal Library at The Hague is beginning to show many such characteristics. It is typical also that the old national rules for title description, a 61-page book of A5 size, gave general rules only and that most libraries had their own rules, which were more or less in accord with the national rules. Provisional new rules have now been framed, based on ISBD(M) and ISBD(S), but it remains uncertain if and to what extent these will be applied.

Needless to say, in the Dutch situation a more centralized approach will meet with resistance from influential local interests. If reduction of local autonomy is feared, defensive groupings will be mustered; independence and locally approved methods will not easily be surrendered. In Holland a compromise between local and central interests still has to be achieved. This is the reason why an operational DUMC could not hitherto be established. On the other hand most map collections, including the three largest with more than 100.000 maps, have such arrears in cataloguing that they cannot make the extra effort needed to change a new (DUMC) system themselves. In order to solve such problems, the board of the NSIDK decided to apply for a subsidy from the Ministry of Education and Sciences for an implementation-phase from 1979 to 1983. The intention is that during this period the DUMC Bureau assists as a national agency important map collections in the cataloguing of their collections. We use criteria as the general importance of the material on hand, when deciding if a map collection is suitable for this assistance. With the aid of DUMC the map collections will be more easily accessible and we will be able to extend our data base from which other collections can benefit via the Mother-Daughter procedure.

If the subsidy should not be granted, and in times of economy such as these even an optimist must accept this possibility, the project will continue to function because of the co-operative relations now established with (for example) the Royal Library, but it would then remain a modest pilot project and the objective of opening up the Dutch map collections would not be achieved.

Explanation of the Figures

The illustrations show the contents of the normal output as made by DUMC. The output can be given in the form of line-printer-cards as well as in the form of C.O.M. fiches.

Figure 1: shows the main data base's outprint and gives coded tags and related texts. These print-outs are presented to third parties to enable them to apply the Mother-Daughter procedure. The meaning of the tags is:

- 008 - fixed fields:
  containing the last up-dated date, followed by the copyright date. NE = The Netherlands as country of publication; D = No Meridian; A = Single Map; 1 = with Index; DUT = in Dutch language; NEGDH = Map Publisher Code.
- 041 - Code for language
- 090 - Local call number, in this case out of the Voluntary Depository
- 110 - Primary responsibility
- 245 - Title
- 260 - Imprint
- 300 - Collation
- 500 - Annotation
- 501 - Price
- 505 - Contents note
- 507 - Scale
- 900 - Co-ordinates
- 910 - Retrievable scale
- 930 - Content date
- 940 - U.D.C. code for region, in this case Noord-Holland
- 941 - U.D.C. code for subject, in this case a city plan

950 - Keywords for region, specifying the U.D.C. region code.

The asterisks denote capitals.

At the moment we are changing to a system in which the input occurs via a Terminal TEXAS Instruments Silent 742. Once this is accomplished, we will have a full ASCII character set at our disposal and will be able to use capitals and small letters in our direct input.

The coding of the tags will be adapted to the prescriptions of UNIMARC.

Figure 2: shows a standard catalogue card with a region code as its entry and sorted by subject, code and scale.

Figure 3: an added entry for the second region keyword.

Figures 4, 5 and 6: have *subject* code, *title* and *primary responsibility* for an entry.

Fig. 1: Uitslag van het hoofdbestand

FILENO.	TAG	SEQ	IND	SUBF	TEXT
D7700434	001				/DUMC
D7700434	008				770804C19751975NE DA 1DUT NEGDH
D7700434	041				DUT.
D7700434	090		0		*K*M492.62 -2540/001/00000/00/1975/1
D7700434	110		2 0	A	[*OPENBARE *WERKEN *DEN *HELDER].
D7700434	245		1 0	A	*DEN *HELDER.
D7700434	245		1 0	F	[*GETEKEND DOOR *OPENBARE *WERKEN *DEN *HELDER].
D7700434	260		0	A	*DEN *HELDER,
D7700434	260		0	B	*GEMEENTE *DEN *HELDER,
D7700434	260		0	C	1975.
D7700434	300			A	1 GEKL. KRT.,
D7700434	300			C	44X63 CM., GEV. 23X16 CM..
D7700434	500			A	*VERSO: STRAATNAMENLIJST *DEN *HELDER EN LIJST MET OPENBARE GEBOUWEN.
D7700434	501				*PRIJS: *FL. 1,50.
D7700434	505		2		*VERSO: OVERZICHTSKAART GEMEENTE *DEN *HELDER, SCHAAL CA. 1 : 50.000 EN PLATTEGR
D7700434	505		2		ONDEN VAN *JULIANADORP, *DROOGHE *WEERT EN *DE *ZANDLOPER, ALLEN SCHAAL CA. 1 :
D7700434	505		2		8.000.
D7700434	507			A	*SCHAAL 1 : 10.000.
D7700434	900			A	004 40'00" *O.*L.
D7700434	900			B	004 50'00" *O.*L.
D7700434	900			C	052 59'00" *N.*B.
D7700434	900			D	052 56'00" *N.*B.
D7700434	910				000010000.
D7700434	930				+1950/+1999.
D7700434	940				(492.62)
D7700434	941				911.375.53
D7700434	950	0	0		*DEN *HELDER
D7700434	950	1	0		*JULIANADORP

Fig. 2: Fiche-regio 1

KM492.62 -2540/001/00000/00/1975/1

(492.62) Den Helder, 911.375.53, 000010000.

DEN Helder. [Getekend door Openbare Werken Den Helder].

Schaal 1 : 10.000.

[Openbare Werken Den Helder].

Den Helder, Gemeente Den Helder, 1975

1 gekl. krt., 44x63 cm., gev. 23x16 cm..

Verso: overzichtskaart gemeente Den Helder, schaal ca. 1 : 50.000 en plattegronden van Julianadorp, Drooghe Weert en De Zandloper, allen schaal ca. 1 : 8.000.
Verso: straatnamenlijst Den Helder en lijst met openbare gebouwen.

Fig. 3: Fiche-regio 2

KM492.62 -2540/001/00000/00/1975/1

(492.62) Julianadorp, 911.375.53, 000010000.

DEN Helder. [Getekend door Openbare Werken Den Helder].

Schaal 1 : 10.000.

[Openbare Werken Den Helder].

Den Helder, Gemeente Den Helder, 1975

1 gekl. krt., 44x63 cm., gev. 23x16 cm..

Verso: overzichtskaart gemeente Den Helder, schaal ca. 1 : 50.000 en plattegronden van Julianadorp, Drooghe Weert en De Zandloper, allen schaal ca. 1 : 8.000.
Verso: straatnamenlijst Den Helder en lijst met openbare gebouwen.

D7700434

Fig. 4: Fiche-thema 1

KM492.62 -2540/001/00000/00/1975/1

911.375.53, (492.62) Den Helder Julianadorp, 000010000.

DEN Helder. [Getekend door Openbare Werken Den Helder].

Schaal 1 : 10.000.

[Openbare Werken Den Helder].

Den Helder, Gemeente Den Helder, 1975.

1 gekl. krt., 44x63 cm., gev. 23x16 cm..

Verso: overzichtskaart gemeente Den Helder, schaal ca. 1 : 50.000 en plattegronden van Julianadorp, Drooghe Weert en De Zandloper, allen schaal ca. 1 : 8.000.
Verso: straatnamenlijst Den Helder en lijst met openbare gebouwen.

Fig. 5: Fiche-titel 1

KM492.62 -2540/001/00000/00/1975/1

DEN HELDER.

(492.62) Den Helder Julianadorp, 911.375.53, 000010000.

Den Helder. [Getekend door Openbare Werken Den Helder].

Schaal 1 : 10.000.

[Openbare Werken Den Helder].

Den Helder, Gemeente Den Helder, 1975.

1 gekl. krt., 44x63 cm., gev. 23x16 cm..

Verso: overzichtskaart gemeente Den Helder, schaal ca. 1 : 50.000 en plattegronden van Julianadorp, Drooghe Weert en De Zandloper, allen schaal ca. 1 : 8.000.
Verso: straatnamenlijst Den Helder en lijst met openbare gebouwen.

Fig. 6: Fiche-auteur 1

KM492.62 -2540/001/00000/00/1975/1

[OPENBARE WERKEN DEN HELDER].

(492.62) Den Helder Julianadorp, 911.375.53,
000010000.

Den Helder. [Getekend door Openbare Werken Den
Helder].

Schaal 1 : 10.000.

[Openbare werken Den Helder].

Den Helder, Gemeente Den Helder, 1975.

1 gekl. krt., 44x63 cm., gev. 23x16 cm..

Verso: overzichtskaart gemeente Den Helder, schaal
ca. 1 : 50.000 en plattegronden van Julianadorp,
Drooghe Weert en De Zandloper, allen schaal ca. 1 :
8.000.
Verso: straatnamenlijst Den Helder en lijst met
openbare gebouwen.

DIE THEMATISCHE KARTE
IHRE KATALOGISIERUNG UND DOKUMENTATION

Von Emil Meynen

Die Thematische Karte, oft auch kurz Themakarte genannt, ist eine Karte, in der Erscheinungen und Sachverhalte zur Erkenntnis um ihrer selbst willen dargestellt sind. Der topographische Unterdruck, Gewässer, Siedlungen, Wegenetz und/oder politische oder administrative Grenzen dienen zur allgemeinen Orientierung des Kartengegenstandes oder zur Einbettung des Themas in die Situation.

Während die Zitierungsform der Topographischen Karte im allgemeinen eine eindeutige Aussage vermittelt, ist gleiches von der Thematischen Karte, vom Einzelfall abgesehen, nicht zu sagen.

Zur großen Zahl der Thematischen Karten tritt die Vielzahl der Themen und der sehr unterschiedliche Zweck der Herstellung. Die Themakarte einer einmaligen Orientierung oder die Karte nur als erläuternde Orientierung zu einem Text ist natürlich von wissenschaftlichen Karten zu scheiden. Nur letztere sind für die wissenschaftliche Dokumentation von Bedeutung. Dies ist für die meisten Karten leicht zu entscheiden, jedoch im Einzelfall auch wieder schwierig. In manchen Fällen wird entscheidend sein, ob die Thematische Karte der Teil einer amtlichen Dokumentation oder eines wissenschaftlichen Werkes ist.

Die wissenschaftliche Themakarte kann sowohl selbständige Karte, sei es Handkarte oder Wandkarte, sein, als auch Beilage oder eingeheftetes Blatt zu einer Veröffentlichung, und schließlich tritt die Thematische Karte auch als einfache Textkarte auf. Die äußere Form sagt in vielen Fällen nichts über den originalen Wert einer Themakarte aus.

Die Katalogisierung einer Sammlung wird meist nur die selbständige Karte berücksichtigen oder gegebenenfalls zusätzlich auch die losen Kartenbeilagen von Veröffentlichungen, sei es ein Buch oder eine Zeitschrift. Eine wissenschaftliche Dokumentation muß jedoch heute, wo viele Forschungsergebnisse außer in selbständigen Themakarten auch in Textkarten veranschaulicht werden, unbedingt diese mit einbeziehen, gleichgültig, ob diese in mehreren Farben oder in Schwarzdruck gegeben sind. Die saubere Ausführung ist schon eher ein Zeichen des Wertes einer Karte.

Die Zitierungsgrundregeln sind bei der Topographischen Karte und der Thematischen Karte gleich: Verfasser, Herausgeber und/oder Veranlasser, Titel, Maßstab, Erscheinungsort mit Verlegerangabe und Erscheinungsdatum, die Größe der Karte, (bzw. der Kartenfläche), und gegebenfalls Angabe der Farben. Bei einer Buch- und Zeitschriftenkarte kommt die Veröffentlichungsstelle hinzu und eine Angabe, ob Textkarte oder Beilage mit Seitenangabe und/oder bei einer Kartenfolge die Kartennummer. Über die Reihenfolge der Angaben mag man unterschiedlicher Auffassung sein. Sie kann sich danach richten, ob die Dokumentation in laufender Textform eines Verzeichnisses oder in Kartenform gegeben wird. Die genannten Grundzitierungsangaben sollten jedoch nicht fehlen.

Die Angaben des Verfassers und auch des Herausgebers oder des Veranlassers sind in den vielen Fällen aufschlußreich. Sie können bereits eine Karte als populäre oder wissenschaftliche Karte kennzeichnen. Namentlich der Herausgeber oder der Veranlasser, wenn er eine staatliche oder administrative Institution ist, geben der Karte ein Gewicht. Zwar ist auch in diesen Fällen zu klären, ob es sich un eine Karte der Werbung oder Tagesunterrichtung oder um die Dokumentation eines wissenschaftlichen Sachverhaltes handelt. Eine wissenschaftliche Dokumentation wird im allgemeinen nur Karten der letzteren Art aufführen.

Die große und stets noch wachsende Zahl der Themakarten zwingt zur Auswahl, zu einer Beschränkung auf Primärkarten, auch Quellenkarten genannt, und auf abgeleitete Themakarten, die zu einer neuen Schau der Zusammenhänge und/oder zu neuen Erkenntnissen führen. Solche Beschränkung ist notwendig, soll der Benutzer nicht in einem unnötigen Dokumentationsbalast versinken. Es gibt für den Benutzer einer Dokumentation Thematischer Karten noch genug Schwierigkeiten, das von ihm Gesuchte zu finden.

Ein Zustand, mit dem sowohl der Bibliothekar wie auch der Dokumentar rechnen muß, ist, daß der Titel einer Thematischen Karte nicht dem Kartengegenstand entspricht oder diesen nur allgemein oder ungenau ausdrückt.

Zur Verzeichnung in einem Standortkatalog einer Bibliothek bzw. der Kartensammlung mögen die allgemeinen Grundangaben einer Kartenzitierung genügen, aber die wissenschaftliche Dokumentation - und auch der Sachkatalog einer Sammlung - verlangt im Falle der Thematischen Karten zur Nachweisung des genauen Kartengegenstandes vielfach eine zusätzliche Erläuterungsangabe.

Ein Beispiel sei gegeben: Der Kartentitel lautet "Der Flugverkehr Frankreichs". Aus einem solchen Titel geht nicht hervor ob die Fluglinien mit oder ohne Flughäfen dargestellt sind. Die Karte mit einem derartigen Titel könnte ebensogut eine Karte des Verkehrsvolumens sein. Zeigt die Karte alle Flughäfen oder nur die Fluglinien mit Hervorhebung der interkontinentalen Linien auf? All diese unterschiedlichen Informationen können sich hinter dem genannten Titel verbergen. Der nach einer bestimmten Karte Suchende wird in solchem Falle eine zusätzliche Bemerkung, zumindest eine Kurzangabe als wünschenswert erachten, die den wirklichen Kartengegenstand nennt; sie sei in Klammern [...] der Zitierung angefügt.

Ähnliches gilt für einen Titel wie "Die Geologie Nigeriens". Der Titel sagt nicht aus, ob die Karte eine geologische Formationskarte oder eine Gesteinskarte mit oder ohne tektonische Lineamente ist, oder ob eine Deckschichtenkarte des anstehenden Gesteins oder eine abgedeckte Karte, die Darstellung eines tieferen Untergrundes vorliegt. Hier würde ebenfalls im Einzelfall ein klärender Zusatz des Dokumentars begrüßt werden.

Bei einem Titel "Die Böden Indiens" müßte in gleicher Weise ein Hinweis gegeben werden, ob es sich um eine Karte der Bodenarten nach der Zusammensetzung des mineralischen Anteils oder nur um eine Karte der Bodentypen, d.h. eine Darstellung der Zustandsstufe der Entwicklung der Bodenprofile handelt; da je nachdem der Kartengegenstand ja ein ganz anderer ist.

"Die Bahnlinien der Sowjetunion", auch ein solcher Titel sagt wenig aus. Ist es eine vollständige Wiedergabe aller Bahnlinien? Sind nur die Fern- und Transkontinentalen Bahnlinien und ihre Haltestellen aufgezeigt? Eine erläuternde Kurzangabe sollte den Benutzer der Dokumentation unterrichten, z.B. zu einer Karte "Die Straßen Dänemarks" [unterscheidet: Verbindungswege, Landstraßen, Autofernstraßen].

Der wissenschaftlich geschulte Kartendokumentar wird jeweils unterscheiden, ob es sich bei einer Themakarte um eine Kartierung, d.h. um die lagegetreuen Aufnahme von kartographisch vermessenen oder durch andere Methoden ermittelte und aufgenommene Erscheinungen oder Sachverhalte handelt, oder um eine Kartographierung, d.h. eine kartographische Zeichnung, die eine zwar lagetreue, aber generalisierte Darstellung von Kartengegenständen aufgrund einer Stoffaufbearbeitung ist. Der Maßstab der Karte läßt erstere erkennen; letztere ist zumeist in kleinerem Maßstab gegeben. Soweit Titel und Maßstab bereits die Kartierung erkennen lassen, z.B. "Bayerische Anstalt für Bodenforschung: Gesteinskarte 1 : 25 000" oder "Pflanzensoziologische Aufnahme 1 : 10 000", ist kein weiterer Zusatz nötig; in anderen Fällen wird der Benutzer des Dokumentationsverzeichnisses für eine Kurzkennzeichnung "[amtliche Kartierung]" oder "[Kartierung 1957]" dankbar sein. Sofern dieser Zusatz fehlt, ist die Annahme berechtigt, daß es sich um eine kartographierte Darstellung handelt.

Sehr häufig ist aus dem Kartentitel z.B. einer kartographierten Themakarte nicht zu ersehen, ob es sich um eine Originalvorlage eines Kartenthemas handelt oder ob wir es mit der Reproduktion eines an anderer Stelle veröffentlichen Originals - möglicherweise auch eines anderen Verfassers - zu tun haben. Der Vermerk auf der Karte "Entwurf: N.N." bedeutet noch nicht immer die Originalität des dargestellten Forschungsergebnisses, sondern im Einzelfall nur eine Umzeichnung einer Originalvorlage. Indes dürfen wir im allgemeinen davon ausgehen, daß der Wissenschaftler bei einer von ihm gezeichneten Karte, sofern der Kartengegenstand aus einer anderen Arbeit stammt, die Herkunft oder Quelle seiner Darstellung nennt. Eine zusätzliche Kurzangabe dieser Grundlage z.B. "[Quelle: E. Ritter, 1974]" oder "[nach geowiss. Forschungsanstalt, 1903]" kann für den Benutzer ein wertvoller Hinweis sein, ob er die Karte einsehen muß oder nicht. Falls der Verfasser die Quelle zitiert, würde die Angabe in spitzen Klammer <...> zu stellen sein.

Wertvoll ist es bei Thematischen Karten zu erfahren, ob eine *Positionskarte* (position map), auch Ortslagekarte bezeichnet (Fig. 1), vorliegt, in der auf einen Punkt bezogene oder in der Karte nur punkthaft dargestellte Erscheinungen oder Sachverhalte mittels Ortslagekartenzeichen dargestellt sind, oder eine *Gebietslagekarte* (position map of an area or areas) (Fig. 2), von E. Imhof auch *Mosaikkarte* genannt, welche die Gebiete als begrenzte Einheiten in ihrer Grundrißfläche gegebenenfalls mit Angabe ihrer Art aufzeigt, oder ob es sich um eine *Arealkarte* (distribution area map) (Fig. 3) handelt, bei der die Grenzlinien nicht in Vermessung gewonnene Grundrißlinien sind, sondern Umgrenzungslinien von Arealen der Verbreitung, z.B. von Biota, meist von sehr subjektiver Linienführung. Der Zusatz "Positionskarte", "Gebietskarte" und "Arealkarte" kann in lapidarer Kürze den Aussagewert einer Karte eindeutig vermitteln.

In weiteren Fällen wird es für den Benutzer eines Katalogs oder Dokumentationsverzeichnisses vorteilhaft sein, zu erkennen, ob es sich bei einer betreffenden Themakarte um eine *Zeitpunktkarte*, die den Kartengegenstand zu einem bestimmten Datum darstellt, handelt, oder um eine *Entwicklungskarte* der Darstellung des Kartengegenstandes, z.B. das Ausbreitungsgebiet der Cholera in Bangladesh über einen längeren Zeitabschnitt mit zu verschiedenen Zeitpunkten geändertem *Grenzenbild*. Es mag des weiteren unterschieden werden eine *Gebietsentwicklungskarte*, die die Flächenentwicklung eines Territorialgebietes zeigt, und eine *Grenzenentwicklungskarte*, die einen Grenzabschnitt in Phasen seiner Vor- und Zurückverlegung veranschaulicht.

Bei einer Darstellung meteorologischer Erscheinungen oder Sachverhalte wäre für die Bewertung der Karte - sofern nicht im Titel ausgedrückt - ein zusätzlicher Dokumentationsvermerk über die Beobachtungszeit, auf der die Darstellung basiert, nützlich, bei einer Karte "Häufigkeit der Gewitter in Süddeutschland" z.B. der Vermerk "aufgrund 30jähriger Beobachtung".

Bei Darstellungen von Kontinua, Darstellungen von Wertefeldern, sei es geophysikalischer, geochemischer oder auch anderer naturgegebener Art, mittels gleichbleibender Intensitäts- oder Wertstufenlinien, sogenannten *Isolinienkarten*

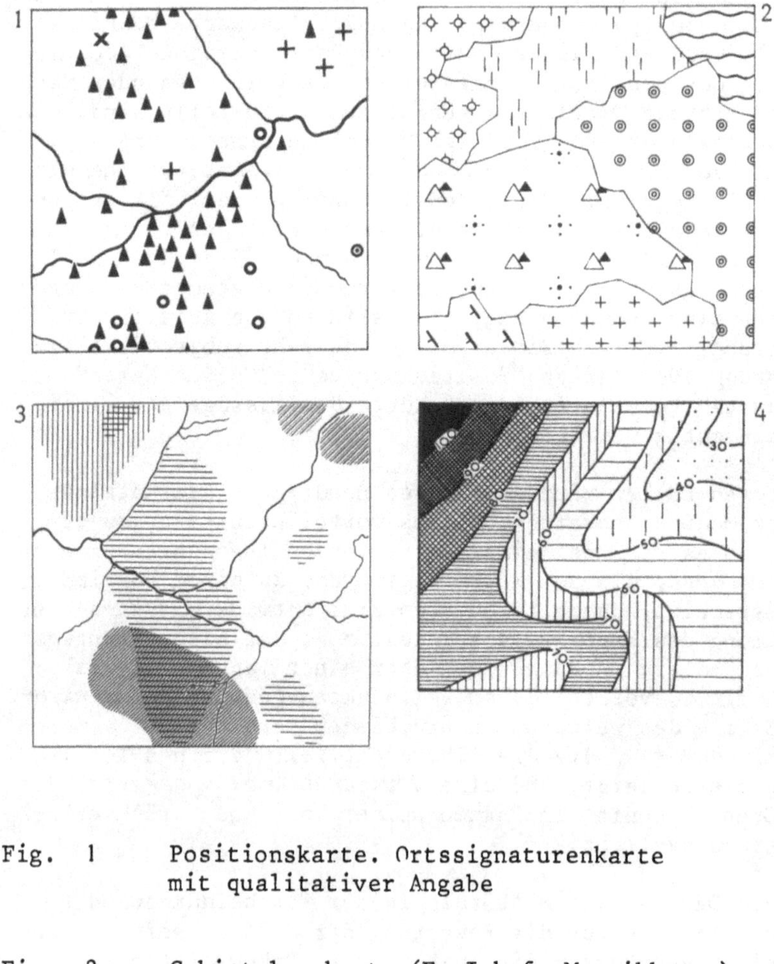

Fig. 1 Positionskarte. Ortssignaturenkarte mit qualitativer Angabe

Fig. 2 Gebietslagekarte (E. Imhof: Mosaikkarte)

Fig. 3 Flächig angelegte Arealkarte

Fig. 4 Isolinienkarte eines naturgegebenen Wertefeldes (Kontinuum)

(isoline map) (Fig. 4), bedarf es ebenfalls eines Zusatzes, wenn der Charakter der Karte nicht aus dem Titel zu entnehmen ist; eine Angabe der Wertstufung solcher "Isolinienkarten" ist jedoch im Einzelfalle sehr aufschlußreich, z.B. "[Isolinienkarte in 25er Stufung]" oder "[Isolinien in 10, 25, 30 und >30]".

Neben naturgegebenen Wertefelddarstellungen finden sich *Isolinienkarten geometrischer Kontinua* z.B. für Angaben der Distanz. Ihr ineinander geschachteltes Liniensystem kann z.B. bei einer "Entfernungskarte der Bahnstationen eines Bahnstreckennetzes" das Linienbild einer *Isodistanzenkarte* mit 1 km Abstand der Isodistanzen sein, oder einer *Isochronenkarte* der Zeitangabe mit z.B. Isolinien von 5, 10, 15 Stundenentfernung. *Elevations(iso)linien* von Schießweiten wie Äquiformatenkarten, letztere Darstellung gleicher Deformation eines Kartennetzes mittels Isolinien, hier Äquiformaten, sind andere Beispiele.

Namentlich die statistische Karte verlangt, sofern der Titel vage ist, vom Dokumentar Zusätze. Den Benutzer interessiert, ob die statistische Themakarte, was häufig nicht im Titel enthalten ist, eine absolute oder eine relative Darstellung ist.

Bei einer statistischen Karte absoluten Aussagenwertes ist des weiteren zu unterscheiden eine Karte mit *formgleichen Größekartenzeichen* (Fig. 5), die nach Größe des Wertes proportional veränderlich sind, und eine Karte mit *gestuften Größenkartenzeichen* (Fig. 6), gegebenenfalls zur Kennzeichnung von Art und Funktion zusätzlich in der Form sich ändernder Größezeichen. In letzterem Falle ist, sofern zur Kennzeichnung der Darstellung eine Kurzerläuterung angebracht ist, eine solche Angabe anzufügen, z.B. "in 10er Stufung" oder bei einer Darstellung mit ungleicher Stufung, wie bei einer Karte der Altersgliederung einer Bevölkerung der Zusatz "<in Stufen <1, 1-18, 19-65, > 65 Jahren>".

Noch ein weiteres will beachtet sein. Größekartenzeichen mögen punkthaft darzustellende Erscheinungen veranschaulichen oder auch auf Flächeneinheiten bezogen sein. Im ersteren Falle würde man von einer *Karte mit Positionsgröße(karten)zeichen*

5

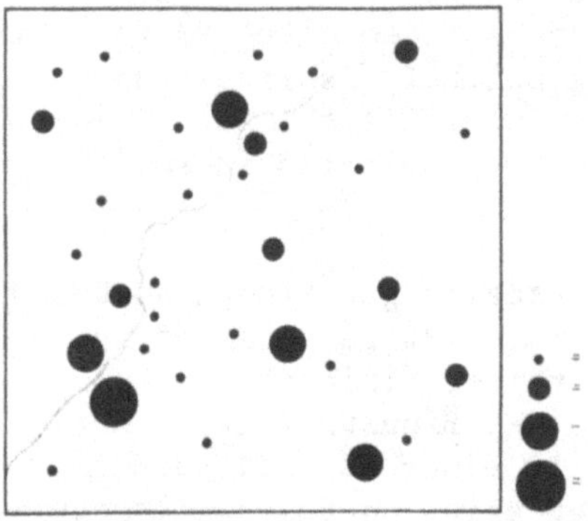

6

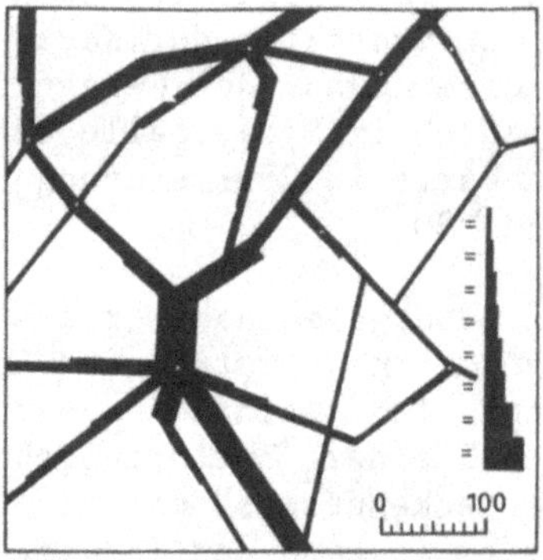

7

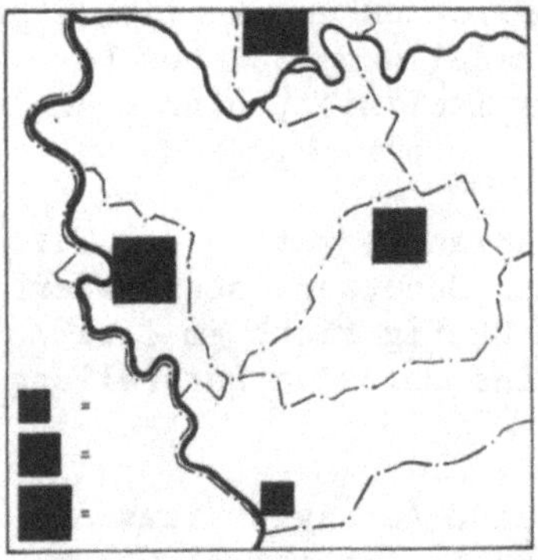

8

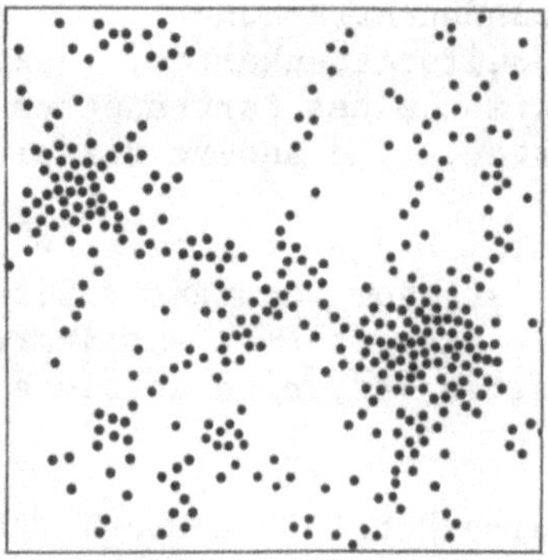

Fig. 5 Statistische Karte: Karte mit formgleichen Größekartenzeichen (Positionsgrößesignaturen)

Fig. 6 Statistische Karte: Karte mit gestuften Linienkartenzeichen (Liniensignaturen)

Fig. 7 Statistische Karte: mit gestuften Gebietskartenzeichen

Fig. 8 Punktstreuungskarte: Streuung der Punkte entsprechend der Siedlungskonzentration

oder Größesignaturen (Fig. 5) sprechen, im zweiten Falle wäre die eindeutige Kennzeichnung *Karte mit gestuften Gebietskartenzeichen* (Fig. 7). Bei Nichtaufzeigung der Gebietsgrenzen würde man von "gestuften Gebietskartenzeichen in Raumlage" sprechen. Das Kartenbild als solches einer Karte mit Positionsgrößekartenzeichen und das einer Karte mit Gebietskartenzeichen in Raumlage unterscheidet sich nicht; um so mehr ist im letzten Fall der Zusatz "in Raumlage" zur Verdeutlichung des Kartengegenstandes klärend.

Zu den absoluten Darstellungen gehört auch die *Punkt(streuungs)-karte* (dot map) (Fig. 8), in der Erscheinungen mittels gleichbleibenden Mengepunkten eingetragen sind. Hier genügt zur Kennzeichnung der Aussage der Karte, sofern nicht vom Verfasser im Titel ausgedrückt, sondern nur in der Legende gegeben, der Zusatz "<Punktstreuungskarte: 1 Punkt = 110>" oder in anderem Falle kurz "<1 Punkt = 10 Personen>".

Außer nach der Menge kann eine Erscheinung auch nach zeitlichen Auftreten und anteiliger Zusammensetzung aufgegliedert dargestellt werden: Es sind die sogenannten *Diagrammkarten* (diagram map) verschiedenster Form. Die Art des Diagramms - Koordinatendiagramm, rechtwinkliges oder Polarkoordinatendiagramm, ein Flächen- oder Körperdiagramm - ist für die Aussage einer Karte unwesentlich; wichtig ist, auf welche Einheit die Darstellung bezogen ist.

Wie im Falle der Kartenzeichenkarte (Signaturenkarte) kann man von einer *Positionsdiagrammkarte* (point diagram map) (Fig. 9), von einer *Liniendiagrammkarte* (line diagram map) (Fig. 10), z.B. Verkehrsvolumen einer Bahnstrecke, und einer *Gebietsdiagrammkarte* (areal diagram map) (Fig. 11) sprechen. Die Zahl dieser Karten absoluter Darstellung ist sehr groß, aber im Titel vielfach nicht immer eindeutig gefaßt; in solchem Falle mag die kurze Begriffsangabe zur eindeutigen Kennzeichnung der Ausführung des Kartenthemas dienen.

Zu den Diagrammkarten gehören auch die *Kurvendiagrammkarten* (Fig. 12), die die Veränderung von Größewerten zu verschiedenen Zeiten, z.B. bei einer Darstellung meteorologischer Erscheinungen monatlich im Jahresverlauf, aufzeigen. Hier könnte, falls nicht im Titel gesagt, die Kurzangabe "Kurvendiagrammkarte in Monatsweiten auf Grund einer 25jährigen Beobachtung" eine klare Ausweisung sein.

9

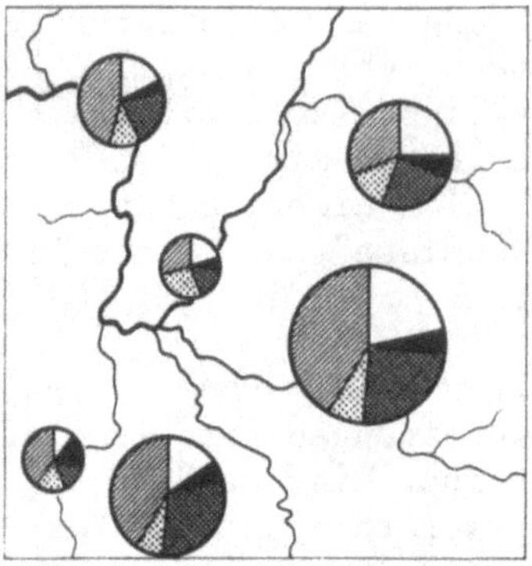

10

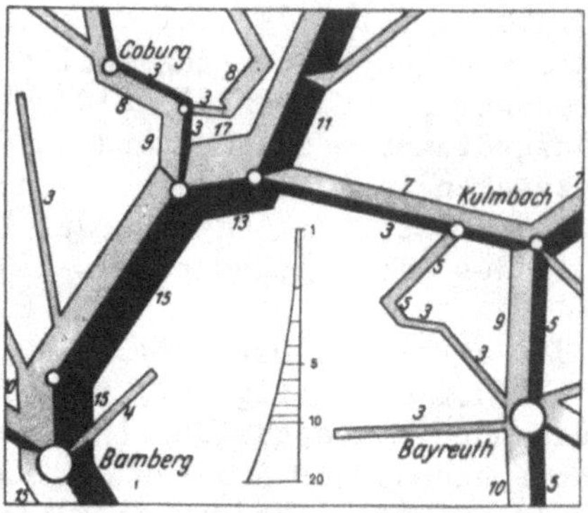

11

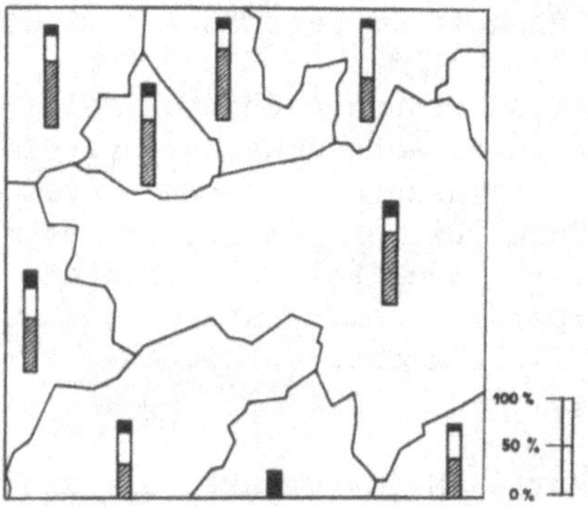

12

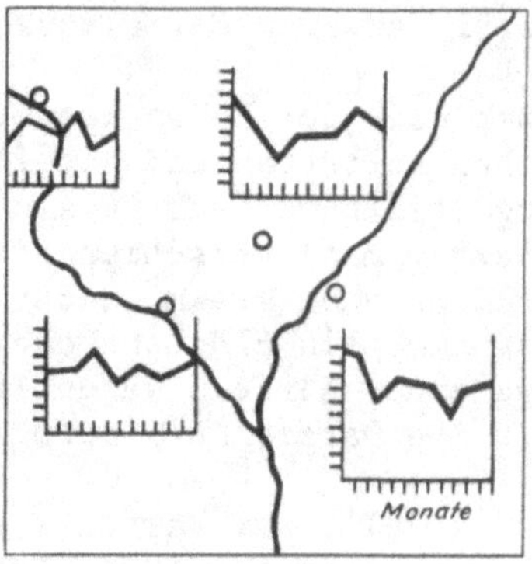

Fig. 9 Positionsdiagrammkarte

Fig. 10 Liniendiagrammkarte

Fig. 11 Gebietsdiagrammkarte

Fig. 12 Kurvendiagrammkarte

Die relative Darstellung liegt in Form von *Gebietsstufenkarten* (Fig. 13) vor, nach englischem Sprachgebrauch auch als *Choroplethenkarten* (choropleth map) bezeichnet. Sie sind die Darstellung von Verhältniswerten für Gebietseinheiten. Statistische Verhältniszahlen als Kartenelemente können sein: Gliederungszahlen (Anteil), z.B. Anteil der Behördenangestellten an der Gesamtzahl der Erwerbstätigen, oder Beziehungszahlen, die entweder Häufigkeits- oder Dichteangabe sind. Eine *Häufigkeitskarte* veranschaulicht einen Sachbezug, z.B. Geburten bezogen auf Frauen im Alter von 20-45 Jahren. Die *Dichtekarten* geben einen Raumbezug an, z.B. "Einwohner auf km^2".

Die Titel im Falle von Gebietsstufenkarten (Choroplethenkarten) geben im allgemeinen genaue Auskunft darüber, was dargestellt ist, und es wird nur in Sonderfällen notwendig sein, eine Zusatzangabe zu machen. Ein Dokumentar sollte sich jedoch der verschiedenen Aussagen bewußt sein, um im Sonderfalle eine zusätzliche Erläuterung geben zu können. Nicht selten geht die Verhältnisaussage nicht aus dem Titel, sondern nur aus der Legende hervor.

Bei der relativen Darstellung kommt es sehr auf die Größe der Gebietseinheiten an, die der Auswertung oder Darstellung zugrunde liegen. Als Gebietseinheiten mögen politische oder administrative Einheiten oder in anderen Fällen geometrische Maschen eines Quadratgitters (Fig. 14) oder Sechseckgitters gewählt sein. Es können auch natur- oder kulturräumliche Einheiten von bestimmter Größenordnung sein. Je größer die Gebietseinheiten sind, die der Darstellung zugrunde liegen, um so mehr wird die örtliche Differenzierung verwischt. Der Dokumentar sollte also, sofern die Angabe nicht im Titel enthalten ist, die zugrunde liegende Gebietseinheit, z.B. "nach Gemeinden", "nach Kreisen" oder die Gebietsgröße "<Mascheneinheit = 1 km^2>" angeben.

Die Kartensprache ist kein ungeregeltes Informationsmittel, sondern in ihrem gesamten Aufbau logisch. Wenn das Ergebnis einer Ausrechnung statistischer Erhebung nach Gebietseinheiten, z.B. Bevölkerungsdichte in jüngerer Zeit über EDV-Verarbeitung mittels Isolinien wiedergegeben wird, so sollte der Dokumentar nicht vorschnell von "Isolinienkarte" sprechen, sondern von einer "Pseudoisolinienkarte" (Fig. 15 b), denn

13

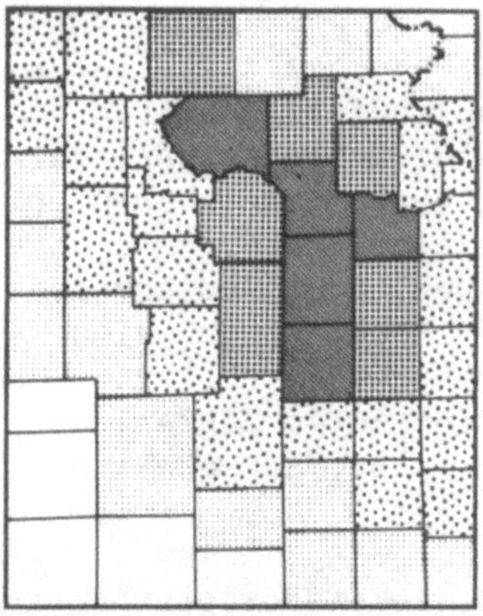

14

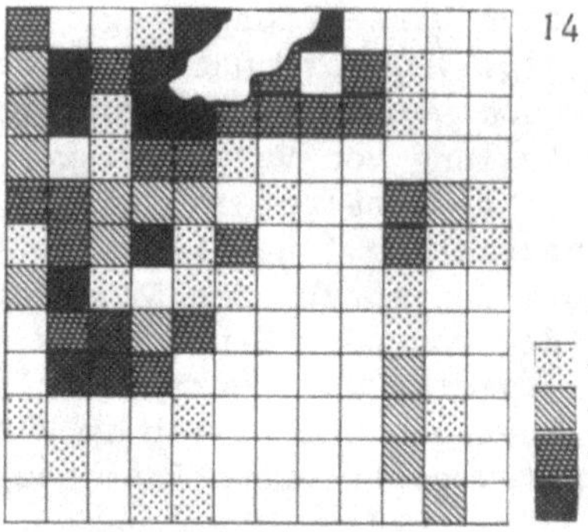

15a 15b

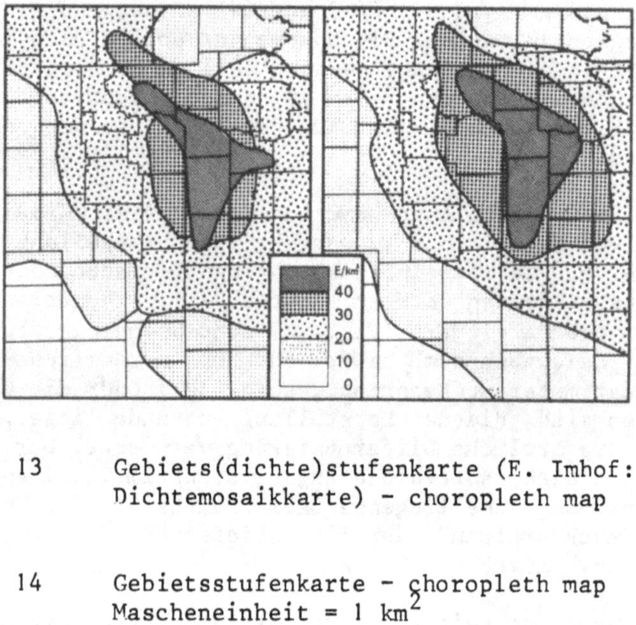

Fig. 13 Gebiets(dichte)stufenkarte (E. Imhof: Dichtemosaikkarte) - choropleth map

Fig. 14 Gebietsstufenkarte - choropleth map Mascheneinheit = 1 km^2

Fig. 15a Generalisierte Gebiets(dichte)stufenkarte

Fig. 15b Pseudoisolinienkarte (Robinson: isarithmic map)

solche Linien gleicher Distanzen verbergen die großen Wertsprünge, die bei einer Dichtekarte, auf Gebietseinheiten bezogen, bestehen. Die Isolinien in diesem Falle sind Täuschung, keine Skalarwertlinien.

Noch zwei andere Kartentypen verlangen eine Erwähnung. Eine Karte mit dem Titel "Der Humboldt-Strom an der Peruküste" kann sowohl als *Vektorkarte* (vector map) (Fig. 16) gegeben sein, die die gemessene Strömungsrichtung durch Vektorzeichen, gleichbleibende Kleinpfeile angibt, die jeweils mit dem Schaftende an der Stelle, wo die Messung oder Beobachtung gemacht wurde, in die Karte eingesetzt sind, als auch eine *Pfeillinienkarte* (arrow map) (Fig. 17), die als abgeleitete Karte mit Pfeillinien den Weg der Strömung generalisiert veranschaulicht. Erst der Zusatz "Vektorkarte", oder "Pfeillinienkarte" läßt den Benutzer einer Kartendokumentation das wirkliche Kartenthema und seine Ausführung wissen.

Ganz allgemein wird der Dokumentar entscheiden, ob es sich bei einer vorliegenden Themakarte um eine *analytische Darstellung* des Kartengegenstandes handelt, die diesen in seiner sachlichen Aufgliederung zeigt, d.i. die dargestellten Erscheinungen oder Sachverhalte werden als Teile eines übergeordneten Formal- oder Funktionsgefüges aufgefaßt, oder ob eine *komplexe Themakarte* vorliegt, in der mehrere Arten von Erscheinungen und/oder Sachverhalten unabhängig voneinander veranschaulicht sind oder um eine *synoptische Karte*, in der Erscheinungen und/oder Sachverhalte aufgezeigt werden, um ihre Zusammenhänge und Entwicklung erkennen zu lassen, oder ob eine *synthetische Themakarte* gegeben ist, in der Raumeinheit aufgrund eines Formal- oder Funktionsgefüges der Erscheinungen und/oder Sachverhalte als synthetische Raumeinheit erfaßt und in ihren Grenzen dargestellt wird. Im allgemeinen wird man bei einer wissenschaftlichen Karte den Titel so gewählt haben, daß der wissenschaftliche Benutzer die Art der Karte sofort erkennt. In anderen Fällen wird ein Zusatz bei der Dokumentationszitierung zur Kennzeichnung angebracht sein, z.B. im Falle einer Karte "Die Einzugsbereiche Rheinland-Pfalz" der Zusatz "Hierarchische Raumgliederung: Unter-, Mittel- und Höhere Bereiche".

Echte Dokumentation, die mehr als bloßes Verzeichnis von Karten sein soll, wird also in vielen Fällen zusätzliche

16
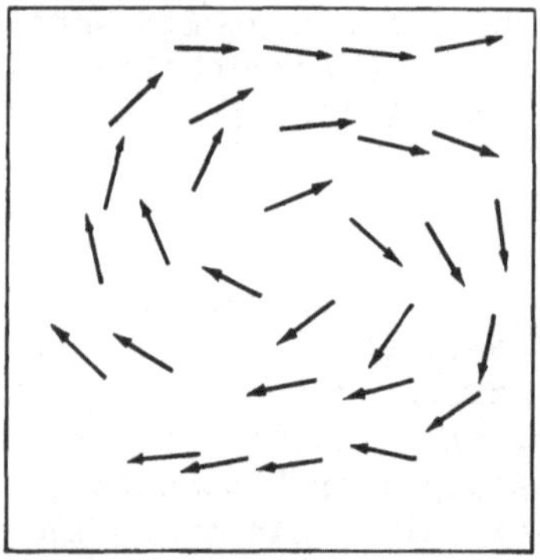
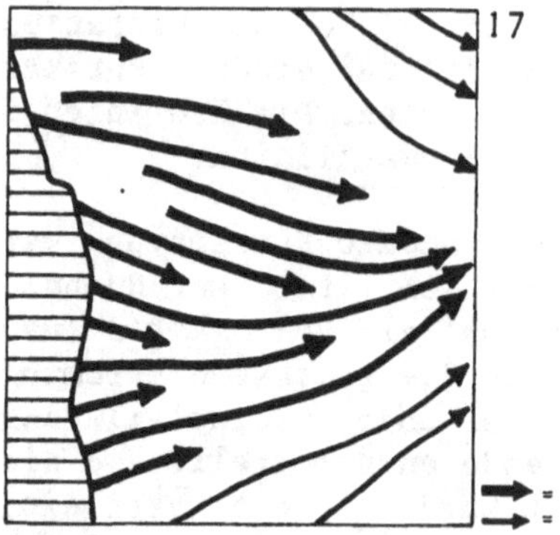
17

Fig. 16 Vektorkarte - vector map

Fig. 17 Pfeillinienkarte - arrow map

Erläuterungen einfügen müssen, wenn der Benutzer in der Suche nach einem bestimmten Kartengegenstand oder einer besonderen Aufschlüsselung des Kartengegenstandes nicht unnütz viele Karten zur Hand nehmen soll. Und dies ist auch in bibliothekarischem Sinne zur Zeitersparung und geringen Abnutzung der Karten wünschenswert. Die Zusätze sollten nicht langatmige Sätze oder gar Wertungen sein, sondern, wie aufgezeigt, formale Kurzangaben.

Neben der Kartendokumentation als solcher, die jeweils die Karte als Einheit dokumentiert, steht die Objektdokumentation von speziellen Fachdokumentationsstellen, die an der Dokumentierung einzelner Objekte des Kartengegenstandes interessiert sind. Z.B. kann eine Dienststelle des öffentlichen Verkehrs interessiert sein, auf welchen Karten die Schnellzugstrecken verzeichnet sind, welche verkehrstechnischen Angaben dargestellt und/oder inwieweit der Verkehr in seiner Funktion oder Leistung veranschaulicht wird, oder ob Planungen eingetragen sind.

Die Objektdokumentation erfolgt heute auf Sichtlochkarten. Hier genügt im allgemeinen die reine Titelnennung mit den Grundangaben (Verfasser, Veranlasser, Maßstab, Datum, Bearbeitungs- bzw. Veröffentlichungsdatum und gegebenenfalls Standortnachweis) als Sichtvermerk. Die genannten Kurzangaben zur Darstellungsform wären wie die Objekte des Kartengegenstandes zu lochen. Hier wird jede Fachdokumentationsstelle ihre eigene Aufgliederung und Anordnung der Lochung wählen.

Neue Wege wirft die EDV-Verarbeitung auf. Doch auch hier wird gerade mit den angedeuteten Kurzangaben, die verschlüsselt werden könnten, die Dokumentation der Thematischen Karte gebrauchsgerechter, d.h. effizienter gestaltet werden können.

CONTRIBUTORS

Carrington, David K.	Head of the Cataloguing Section of the Geography and Map Division, The Library of Congress, Washington, D.C.
De Smet, Antoine, Dr.	Conservateur honoraire et Chef du Département des Cartes et Plans en retraite de la Bibliothèque royale Albert Ier, Bruxelles
Drazniowsky, Roman, Dr.	Map Curator of the American Geographical Society at the Library of the University of Wisconsin, Milwaukee, Wisc.
Kejlbo, Ib Rønne	Chief of the Kort- og Billedafdelingen, Det Kongelige Bibliotek, København; Chairman of the IFLA Subsection Geography and Map Libraries 1970-1977
Kozlowa, Anna V., Dr.	Head of the Map Department, Lenin State Library, Moskva
Mickwitz, Ann Mari (†)	Nordenskiöld Collection of the University Library, Helsinki
Meynen, Emil, Prof.Dr.	Direktor emeritus des Instituts für Landeskunde in der Bundesforschungsanstalt für Landeskunde und Raumordnung, Bonn-Bad Godesberg
Modelski, Andrew M.	Bibliographer at the Geography and Map Division, The Library of Congress, Washington, D.C.

Pognon, Edmond	Conservateur en chef (en retraite) du Département des Cartes et Plans, Bibliothèque nationale, Paris
Stephenson, Richard W.	Head of the Reference and Bibliography Section of the Geography and Map Division, The Library of Congress, Washington, D.C.
Waal, E. Hans van de, Dr.	Director of the Dutch Union Map Catalogue at the Geographical Institute of the Rijksuniversiteit Utrecht; Secretary of the IFLA Section Geography and Map Libraries
Wallis, Helen, Dr.	The Map Librarian of the British Library Reference Division, London; Chairman of the IFLA Section Geography and Map Libraries from 1977, Secretary 1969-1977
Winearls, Joan	Map Librarian at the University of Toronto Library, Toronto
Wolter, John, Dr.	Succeeded W.W. Ristow as Chief of the Geography and Map Division, The Library of Congress, Washington,D.C.
Woodward, David, Dr.	Program Director of the Hermon Dunlap Smith Center for the History of Cartography at the Newberry Library, Chicago, Ill.
Zögner, Lothar, Dr.	Head of the Map Department, Staatsbibliothek Preußischer Kulturbesitz, Berlin (West)

INDEX

A

B

D

E

F

I

J

K

W

Y

Kolibri 360 GmbH
Ferdinand-Jühlke-Straße 7
99095 Erfurt, Deutschland
produktsicherheit@kolibri360.de